Preface

This Engineering Science pocket book is intended to provide students, technicians and engineers with a readily available reference to the essential engineering science formulae, definitions and general information needed during their studies and/or work situation.

The book assumes little previous knowledge, is suitable for a wide range of courses, and will be particularly useful for students studying for Technician certificates and diplomas, and for CSE and 'O' and 'A' levels.

The author would like to express his appreciation for the friendly cooperation and helpful advice given to him by the publishers and by the editor Mr A.J.C. May and also to Mr. D.S. Ayling for his agreeing to the use of some material from his *Mechanical Science Checkbook*. Thanks are also due to Mrs. Elaine Woolley for the excellent typing of the manuscript.

Finally, the author would like to add a word of thanks to his wife Elizabeth for her patience, help and encouragement during the preparation of this book.

J O Bird
Highbury College of Technology
Portsmouth

Contents

1 SI units

1   The system of units used in engineering and science is the
Système Internationale d'Unites (International system of units),
usually abbreviated to SI units, and is based on the metric system.
This was introduced in 1960 and is now adopted by the majority
of countries as the official system of measurement.

2   The basic units in the SI system are given in *Table 1.1*.

Table 1.1

Quantity	Unit
length	metre, m
mass	kilogram, kg
time	second, s
electric current	ampere, A
thermodynamic temperature	kelvin, K
luminous intensity	candela, cd
amount of substance	mole, mol

3   SI units may be made larger or smaller by using **prefixes**
which denote multiplication or division by a particular
amount. The eight most common multiples, with their
meaning, are listed in *Table 1.2*.

Table 1.2

Prefix	Name	Meaning
T	tera	multiply by 1 000 000 000 000 (i.e. $\times 10^{12}$)
G	giga	multiply by 1 000 000 000 (i.e. $\times 10^{9}$)
M	mega	multiply by 1 000 000 (i.e. $\times 10^{6}$)
k	kilo	multiply by 1 000 (i.e. $\times 10^{3}$)
m	milli	divide by 1 000 (i.e. $\times 10^{-3}$)
μ	micro	divide by 1 000 000 (i.e. $\times 10^{-6}$)
n	nano	divide by 1 000 000 000 (i.e. $\times 10^{-9}$)
p	pico	divide by 1 000 000 000 000 (i.e. $\times 10^{-12}$)

4   (i) **Length** is the distance between two points. The standard unit of length is the **metre**, although the **centimetre, cm, millimetre, mm** and **kilometre, km**, are often used.

$$1 \text{ cm} = 10 \text{ mm}; \quad 1 \text{ m} = 100 \text{cm} = 1000 \text{ mm};$$
$$1 \text{ km} = 1000 \text{ m}.$$

(ii) **Area** is a measure of the size or extent of a plane surface and is measured by multiplying a length by a length. If the lengths are in metres then the unit of area is the **square metre, m^2.**

$$1 \text{ m}^2 = 1 \text{ m} \times 1 \text{ m} = 100 \text{ cm} \times 100 \text{ cm} = 10\,000 \text{ cm}^2 \text{ or}$$
$$10^4 \text{ cm}^2$$
$$= 1000 \text{ mm} \times 1000 \text{ mm} = 1\,000\,000 \text{ mm}^2$$
$$\text{or } 10^6 \text{ mm}^2$$

Conversely, $1 \text{ cm}^2 = 10^{-4} \text{ m}^2$ and $1 \text{ mm}^2 = 10^{-6} \text{ m}^2$.

(iii) **Volume** is a measure of the space occupied by a solid and is measured by multiplying a length by a length by a length. If the lengths are in metres then the unit of volume is in **cubic metres, m^3**.

$$1 \text{ m}^3 = 1 \text{ m} \times 1 \text{ m} \times 1 \text{ m} = 100 \text{ cm} \times 100 \text{ cm} \times 100 \text{ cm}$$
$$= 10^6 \text{ cm}^3$$
$$= 1000 \text{ mm} \times 1000 \text{ mm} \times 1000 \text{ mm} = 10^9 \text{ mm}^3$$

Conversely, $1 \text{ cm}^3 = 10^{-6} \text{ m}^3$ and $1 \text{ mm}^3 = 10^{-9} \text{ m}^3$

Another unit used to measure volume, particularly with liquids, is the litre (l) where 1 litre $= 1000 \text{ cm}^3$.

(iv) **Mass** is the amount of matter in a body and is measured in **kilograms, kg**.

$$1 \text{ kg} = 1000 \text{ g} \text{ (or conversely, } 1 \text{ g} = 10^{-3} \text{ kg}) \text{ and}$$
$$1 \text{ tonne (t)} = 1000 \text{ kg}.$$

5   **Derived SI units** use combinations of basic units and there are many of them. Two examples are:

velocity — metres per second, (m/s)
acceleration — metres per second square, (m/s^2).

(a) The unit of **charge** is the coulomb, (C), where one coulomb is one ampere second. (1 coulomb $= 6.24 \times 10^{18}$ electrons). The coulomb is defined as the quantity of electricity which flows past a given point in an electric circuit when a current of one ampere is maintained for one second. Thus charge in coulombs, $Q = It$

where I is the current in amperes and t is the time in seconds.

(b) The unit of **force** is the newton, (N), where one newton is one kilogram metre per second squared. The newton is defined as the force which, when applied to a mass of one kilogram, gives it an acceleration of one metre per second squared. Thus

force in newtons, $F = ma$,

where m is the mass in kilograms and a is the acceleration in metres per second squared. Gravitational force, or weight, is mg, where $g = 9.81$ m/s^2.

(c) The unit of **work or energy** is the joule, (J), where one joule is one newton metre. The joule is defined as the work done or energy transferred when a force of one newton is exerted through a distance of one metre in the direction of the force. Thus

work done on a body in joules, $W = Fs$,

where F is the force in newtons and s is the distance in metres moved by the body in the direction of the force. Energy is the capacity for doing work.

(d) (i) The unit of **power** is the watt, (W), where one watt is one joule per second. Power is defined as the rate of doing work or transferring energy. Thus:

power in watts, $P = \dfrac{W}{t}$,

where W is the work done or energy transferred in joules and t is the time in seconds.

Hence, energy in joules, $W = Pt$.

(e) The unit of **electric potential** is the volt (V) where one volt is one joule per coulomb. One volt is defined as the difference in potential between two points in a conductor which, when carrying a current of one ampere dissipates a power of one watt.

$$\left(\text{i.e. volts} = \frac{\text{watts}}{\text{amperes}} = \frac{\text{joules/second}}{\text{amperes}} = \frac{\text{joules}}{\text{ampere seconds}}\right.$$

$$\left. = \frac{\text{joules}}{\text{coulomb}}\right).$$

A change in electric potential between two points in an electric circuit is called a potential difference. The electromotive force (e.m.f.) provided by a source of energy such as a battery or a generator is measured in volts.

2 **Density**

1   (i) **Density** is the mass per unit volume of a substance. The symbol used for density is ρ (Greek letter rho) and its units are kg/m^3.

$$\textbf{Density} = \frac{\textbf{mass}}{\textbf{volume}}, \text{ i.e.,}$$

$$\boxed{\rho = \frac{m}{v}} \quad \text{or} \quad \boxed{m = \rho V} \quad \text{or} \quad \boxed{V = \frac{m}{\rho}}$$

where m is the mass in kg, V is the volume in m^3 and ρ is the density in kg/m^3.

(ii) Some typical values of densities include:

aluminium $2\,700$ kg/m^3,	copper $8\,900$ kg/m^3,
lead $11\,400$ kg/m^3,	cast iron $7\,000$ kg/m^3,
steel $7\,800$ kg/m^3,	water $1\,000$ kg/m^3,
cork 250 kg/m^3,	petrol 700 kg/m^3.

2   (i) The **relative density** of a substance is the ratio of the density of the substance to the density of water,

$$\text{i.e. relative density} = \frac{\text{density of substance}}{\text{density of water}}$$

Relative density has no units, since it is the ratio of two similar quantities.

(ii) Typical values of relative densities can be determined from para. 1, (since water has a density of 1000 kg/m^3), and include: aluminium 2.7, copper 8.9, lead 11.4, cast iron 7.0, steel 7.8, cork 0.25, petrol 0.7.

(iii) The relative density of a liquid (formerly called the 'specific gravity') may be measured using a **hydrometer**.

3 **Atomic structure of matter**

1   There is a very large number of different substances in existence, each substance containing one or more of a number of basic materials called elements. *'An element is a substance which cannot be separated into anything simpler by chemical means.'*

There are 92 naturally occurring elements and 13 others which have been artificially produced. Some examples of common elements with their symbols are:

Hydrogen H   Helium He,
Carbon C,   Nitrogen N,
Oxygen O,   Sodium Na,
Magnesium Mg,   Aluminium Al,
Silicon Si,   Phosphorus P,
Sulphur S,   Potassium K,
Calcium Ca,   Iron Fe,
Nickel Ni,   Copper Cu,
Zinc Zn,   Silver Ag,
Tin Sn,   Gold Au
Mercury Hg,   Lead Pb
Uranium U.

2   Elements are made up of very small parts called atoms.

> *'An atom is the smallest part of an element which can take part in a chemical change and which retains the properties of the element.'*

Each of the elements has a unique type of atom.

In atomic theory, a model of an atom can be regarded as a miniature solar system. It consists of a central nucleus around which negatively charged particles called electrons orbit in certain fixed bands called shells. The nucleus contains positively charged particles called protons and particles having no electrical charge called neutrons. An electron has a very small mass compared with protons and neutron. An atom is electrically neutral, containing the same number of protons as electrons. The number of protons in an atom is called the **atomic number** of the element of which the atom is part. The arrangement of the elements in order of their atomic number is known as the **periodic table**.

The simplest atom is hydrogen which has 1 electron orbiting

the nucleus and 1 proton in the nucleus. The atomic number of hydrogen is thus 1. The hydrogen atom is shown diagrammatically in *Figure 3.1(a)*. Helium has 2 electrons orbiting the nucleus, both of them occupying the same shell at the same distance from the nucleus, as shown in *Figure 3.1(b)*.

The first shell of an atom can have up to 2 electrons only, the second shell can have up to 8 electrons only and the third shell up to 18 electrons only. Thus an aluminium atom which has 13 electrons orbiting the nucleus is arranged as shown in *Figure 3.1(c)*.

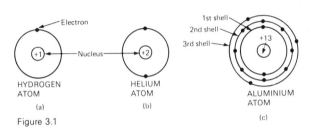

HYDROGEN ATOM

(a)

HELIUM ATOM

(b)

ALUMINIUM ATOM

(c)

Figure 3.1

3   When elements combine together, the atoms join to form a basic unit of a new substance. This independant group of atoms bonded together is called a molecule.

'*A molecule is the smallest part of a substance which can have a separate stable existence.*'

All molecules of the same substance are identical.

Atoms and molecules are the **basic building blocks** from which matter is constructed.

4   When elements combine chemically their atoms interlink to form molecules of a new substance called a compound.

'*A compound is a new substance containing two or more elements chemically combined so that their properties are changed.*'

For example, the elements hydrogen and oxygen are quite unlike water, which is the compound they produce when chemically combined.

The components of a compound are in fixed proportion and are difficult to separate.

Examples of compounds include:
    (i) water H_2O, where 1 molecule is formed by 2 hydrogen atoms combining with 1 oxygen atom,
    (ii) carbon dioxide, CO_2, where 1 molecule is formed by

7

1 carbon atom combining with 2 oxygen atoms,
(iii) sodium chloride NaCl (common salt), where 1
molecule is formed by 1 sodium atom combining with 1
chlorine atom, and
(iv) copper sulphate, $CuSO_4$, where 1 molecule is formed
by 1 copper atom, 1 sulphur atom and 4 oxygen atoms
combining.

5   *'A mixture is a combination of substances which are not chemically
joined together.'*

Mixtures have the same properties as their components. Also, the
components of a mixture have no fixed proportion and are easy to
separate.

Examples of mixtures include:
        (i)   oil and water;
        (ii) sugar and salt;
        (iii) air, which is a mixture of oxygen, nitrogen, carbon
              dioxide and other gases;
        (iv) iron and sulphur;
        (v)   sand and water.

6   *'A solution is a liquid in which other substances are dissolved.'*

A solution is a mixture from which the two constituents may not
be separated by leaving it to stand or by filtration. For example,
sugar dissolves in tea, salt dissolves in water and copper sulphate
crystals dissolve in water leaving it a clear blue colour. The
substance which is dissolved, which may be solid, liquid or gas, is
called the **solute**, and the liquid in which it dissolves is called the
solvent.

    Hence, **solvent + solute = solution**.

A solution has a clear appearance and remains unchanged with
time.

7   *'A suspension is a mixture of a liquid and particles of a solid which
do not dissolve in the liquid.'*

The solid may be separated from the liquid by leaving the suspension to stand or by filtration.

Examples of suspensions include:
        (i)   sand in water,
        (ii) chalk in water,
        (iii) petrol and water.

8   (i) If a material dissolves in a liquid the material is said
        to be **soluble**. For example, sugar and salt are both
        soluble in water.

(ii) If, at a particular temperature, sugar is continually added to water and the mixture stirred there comes a point when no more sugar can dissolve. Such a solution is called saturated.

'A solution is saturated if no more solute can be made to dissolve, with the temperature remaining constant.'

(iii) *'Solubility is a measure of the maximum amount of a solute which can be dissolved in 0.1 kg of a solvent, at a given temperature.'*

For example, the solubility of potassium chloride at 20°C is 34 g per 0.1 kg of water, or, its percentage solubility is 34%.

(iv) The temperature of a mixture, the size of particles of the solute and the agitation of the mixture are factors which influence the solubility of a solid in a liquid.

9   A **crystal** is a regular, orderly arrangement of atoms or molecules forming a distinct pattern, i.e. an orderly packing of basic building blocks of matter.

Most solids are crystalline in form and these include crystals such as common salt and sugar as well as the metals. Substances which are non-crystalline are called **amorphous**, examples include glass and wood.

Crystallization is the process of isolating solids from solution in a crystalline form. This may be carried out by adding a solute to a solvent until saturation is reached, raising the temperature, adding more solute and repeating the process until a fairly strong solution is obtained, and then allowing the solution to cool, when crystals will separate.

There are several examples of crystalline form which occur naturally; examples include graphite, quartz, diamond and common salt.

10   Crystals can vary in size but always have a regular geometric shape with flat faces, straight edges and having specific angles between the sides. Two common shapes of crystals are shown in *Figure 3.2*. The angles between the faces of the common salt crystal (*Figure 3.2(a)*) are always 90° and those of a quartz crystal (*Figure 3.2(b)*) are always 60°. A particular material always produces exactly the same shape of crystal. *Figure 3.3* shows a crystal lattice of sodium chloride. This is always a cubic shaped crystal being made up of 4 sodium atoms and 4 chlorine atoms. The sodium chloride crystals then join together as shown.

11   Metals are **polycrystalline** substances. This means that they are made up of a large number of crystals joined at the boun-

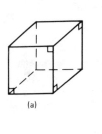 

(a)

(b)

Figure 3.2

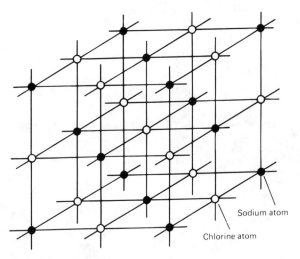

Sodium atom

Chlorine atom

Figure 3.3

daries, the greater the number of boundaries the stronger the material.

12   Every metal, in the solid state, has its own crystal structure. To form an **alloy** different metals are mixed when molten, since in the molten state they do not have a crystal lattice. The molten solution is then left to cool and solidify. The solid formed is a

mixture of different crystals and an alloy is thus referred to as a **solid solution**.

Examples of alloys include:

(i) brass, which is a combination of copper and zinc,

(ii) steel, which is mainly a combination of iron and carbon,

(iii) bronze, which is a combination of copper and tin.

Alloys are produced to enhance the properties of the metal, such as greater strength. For example, when a small proportion of nickel (say 2%–4%) is added to iron the strength of the material is greatly increased. By controlling the percentage of nickel added, materials having different specifications may be produced.

4 Basic chemical reactions

1   A **chemical reaction** is an interaction between substances in which atoms are rearranged. A new substance is always produced in a chemical reaction.

2   **Air** is a mixture, and its composition by volume is approximately: Nitrogen 78%, oxygen 21%, other gases (including carbon dioxide) 1%.

3   If a substance, such as powdered copper, of known mass, is heated in air, allowed to cool, and its mass remeasured, it is found that the substance has gained in mass. This is because the copper has absorbed oxygen from the air and changed into copper oxide. In addition, the proportion of oxygen in the air passed over the copper will decrease by the same amount as the gain in mass by the copper.

4   All materials require the presence of oxygen for burning to take place. Any substance burning in air will combine with the oxygen. This process is called **combustion**, and is an example of a chemical reaction between the burning substance and the oxygen in the air, the reaction producing heat. The chemical reaction is called **oxidation**.

5   An element reacting with oxygen produces a compound which contains only atoms of the original element and atoms of oxygen. Such compounds are called **oxides**. Examples of oxides include: copper oxide CuO, hydrogen oxide H_2O (i.e. water) and carbon dioxide CO_2.

6   **Rusting** of iron (and iron-based materials) is due to the formation on its surface of hydrated oxide of iron produced by a chemical reaction. Rusting of iron always requires the presence of oxygen and water.

7   Any iron or steel structure exposed to moisture is susceptible to rusting. This process, which cannot be reversed, can be dangerous since structures may be weakened by it. Rusting may be prevented by:

    (i) painting with water-resistant paint,
    (ii) galvanising the iron,
    (iii) plating the iron,
    (iv) an oil or grease film on the surface.

8   To represent a reaction a chemical shorthand is used. A symbol represents an element (such as H for hydrogen, O for oxygen, Cu for copper, Zn for zinc, and so on) and a formula represents a compound and gives the type and number of elements in the compound. For example, one molecule of sulphuric acid, H_2SO_4, contains 2 atoms of hydrogen, 1 atom of sulphur and 4 atoms of oxygen. Similarly, a molecule of methane gas, CH_4, contains 1 atom of carbon and 4 atoms of hydrogen.

9   The rearrangement of atoms in a chemical reaction are shown by **chemical equations** using formulae and symbols. For example:

(i) $S + O_2 = SO_2$

i.e. 1 molecule of sulphur, S, added to 1 molecule of oxygen, O_2, causes a reaction and produces 1 molecule of sulphur dioxide, SO_2.

(ii) $Z_n + H_2SO_4 = Z_nSO_4 + H_2$

i.e. 1 molecule of zinc, Z_n, added to 1 molecule of sulphuric acid, H_2SO_4, causes a reaction and produces 1 molecule of zinc sulphate, Z_nSO_4, and 1 molecule of hydrogen, H_2.

10   In a chemical equation:

(i) each element must have the same total number of atoms on each side of the equation. For example, in chemical equation (ii) of para. 9, each side of the equation has 1 zinc atom, 2 hydrogen atoms, 1 sulphur atom and 4 oxygen atoms.

(ii) a number written in front of a molecule multiplies all the atoms in that molecule. For example, the reaction described in para. 3 is:

$$2Cu + O_2 = 2CuO.$$

11   An **acid** is a compound containing hydrogen in which the hydrogen can be easily replaced by a metal. For example, in para. 9, it is shown that zinc reacts with sulphuric acid to give zinc sulphate and hydrogen. An acid produces hydrogen ions H^+ in solution (an ion being a particle formed when atoms or molecules lose or gain electrons).

Examples of acids include:

sulphuric acid, H_2SO_4, hydrochloric acid, HCl, and nitric acid, HNO_3.

12   A **base** is a substance which can neutralise an acid (i.e., remove the acidic properties of acids). An **alkali** is a soluble base. When in solution an alkali produces hydroxyl ions, OH^-

Examples of alkalis include: sodium hydroxide NaOH (i.e., caustic soda), calcium hydroxide, $Ca(OH)_2$, ammonium hydroxide NH_4OH and potassium hydroxide, KOH (caustic potash).

13   A **salt** is the product of the neutralisation between an acid and a base, i.e. acid + base = salt + water.

For example:

$$HCl + NaOH = NaCl + H_2O$$
$$H_2SO_4 + 2KOH = K_2SO_4 + 2H_2O$$
and $$H_2SO_4 + CuO = CuSO_4 + H_2O$$

Examples of salts include: sodium chloride, NaCl (common salt), potassium sulphate, K_2SO_4, copper sulphate. $CuSO_4$ and calcium carbonate, $CaCO_3$ (limestone).

14   An **indicator** is a chemical substance, which when added to a solution, indicates the acidity or alkalinity of the solution by changing colour. Litmus is a simple two-colour indicator which turns red in the presence of acids and blue in the presence of alkalis.

Two other examples of indicators are ethyl orange (red for acids, yellow for alkalis) and phenolphthalein (colourless for acids, pink for alkalis).

15   The **pH scale** (pH meaning 'the potency of hydrogen'), represents, on a scale from 0 to 14, degrees of acidity and alkalinity. 0 is strongly acidic, 7 is neutral and 14 is strongly alkaline.

Some average pH values include: concentrated hydrochloric acid HCl 1.0, lemon juice 3.0, milk 6.6, pure water 7.0, sea water 8.2, concentrated sodium hydroxide NaOH, 13.0.

5 Scalar and vector quantities

Quantities used in engineering and science can be divided into two groups:

(a) **Scalar quantities** have a size or magnitude only and need no other information to specify them. Thus, 10 cm, 50 sec, 7 litres and 3 kg are all examples of scalar quantities.

(b) **Vector quantities** have both a size or magnitude and a direction, called the line of action of the quantity. Thus, a velocity of 50 km/h due east, on acceleration of 9.8 m/s^2 vertically downwards and a force of 15 N at an angle of 30° are all examples of vector quantities.

6 Standard quantity symbols and their units

Quantity	Quantity symbol	Unit	Unit symbol
Acceleration: gravitational	g	metres per second squared	m/s^2
linear	a	metres per second squared	m/s^2
Angular acceleration	α	radians per second squared	rad/s^2
Angular velocity	ω	radians per second	rad/s
Area	A	square metres	m^2
Area, second moment of	I	(metre)^4	m^4
Capacitance	C	farad	F
Capacity	V	litres	l
Coefficient of friction	μ	No unit	
Coefficient of linear expansion	α	per degree Celsius	$/°C$
Conductance	G	seimens	S
Cubical expansion, coefficient of	γ	per degree Celsius	$/°C$
Current	I	ampere	A
Density	ρ	kilogram per cubic metre	kg/m^3
Density, relative	d	no unit	
Dryness fraction	x	no unit	
Efficiency	η	no unit	
Elasticity, modulus of	E	Pascal ($1\ Pa = 1\ N/m^2$)	Pa
Electric field strength	E	volts per metre	V/m

Quantity	Quantity symbol	Unit	Unit symbol
Electric flux density	D	coulomb per square metre	C/m^2
Energy	W	joules	J
Energy, internal	U, E	joules	J
Energy, specific internal	u, e	kilojoules per kilogram	kJ/kg
Enthalpy	H	joules	J
Enthalpy, specific	h	kilojoules per kilogram	kJ/kg
Entropy	S	kilojoules per kelvin	kJ/K
Expansion: coefficient of cubical	γ	per degree Celsius	/°C
coefficient of linear	α	per degree Celsius	/°C
coefficient of superficial	β	per degree Celsius	/°C
Field strength: electric	E	volts per metre	V/m
magnetic	H	ampere per metre	A/m
Flux density: electric	D	coulomb per square metre	C/m^2
magnetic	B	tesla	T
Flux: electric	ψ	coulomb	C
magnetic	Φ	weber	Wb
Force	F	newtons	N
Frequency	f	hertz	Hz
Heat capacity, specific	c	kilojoules per kilogram kelvin	kJ/(kg K)
Impedance	Z	ohm	Ω
Inductance: self	L	henry	H
mutual	M	henry	H
Internal energy	U, E	joules	J
specific	u, e	kilojoules per kilogram	kJ/kg
Inertia, moment of	I, J	kilogram metre squared	kg m^2
Length	l	metre	m

Quantity	Quantity symbol	Unit	Unit symbol
Luminous intensity	I	candela	cd
Magnetic field strength	H	ampere per metre	A/m
Magnetic flux	Φ	weber	Wb
density	B	tesla	T
Magnetomotive force	F	ampere	A
Mass	m	kilogram	kg
Mass, rate of flow	V	cubic metres per second	m^3/s
Modulus of elasticity	E	Pascal	Pa
rigidity	G	Pascal	Pa
Moment of force	M	newton metre	N m
Moment of inertia	I, J	kilogram metre squared	$kg\ m^2$
Mutual inductance	M	henry	H
Number of turns in a welding	N	no unit	
Periodic time	T	second	s
Permeability:			
absolute	μ	henry per metre	H/m
absolute of free space	μ_o	henry per metre	H/m
relative	μ_r	no unit	
Permitivity:			
absolute	ε	farad per metre	F/m
of free space	ε_o	farad per metre	F/m
relative	ε_r	no unit	
Polar moment of area	J	$(metre)^4$	m^4
Power: apparent	S	volt ampere	VA
active	P	watt	W
reactive	Q	volt ampere reactive	VAr
Pressure	p	Pascal ($1\ Pa = 1\ N/m^2$)	Pa
Quantity of heat	Q	joule	J

Quantity	Quantity symbol	Unit	Unit symbol
Quantity of electricity	Q	coulomb	C
Reactance	X	ohm	Ω
Reluctance	S	per henry or ampere per weber	/H or A/Wb
Resistance	R	ohm	Ω
Resistivity	ρ	ohm metre	Ω m
Second moment of area	I	$(\text{metre})^4$	m^4
Shear strain	γ	no unit	
stress	τ	Pascal	Pa
Specific gas constant	R	kilojoules per kilogram kelvin	kJ/(kg K)
Specific heat capacity	c	kilojoules per kilogram kelvin	kJ/(kg K)
Specific volume	v	cubic metres per kilogram	m^3/kg
Strain, direct	ε	no unit	
Stress, direct	σ	Pascal	Pa
Shear modulus of rigidity	G	Pascal	Pa
Temperature coefficient of resistance	α	per degree Celsius	/°C
Temperature, thermodynamic	T	kelvin	K
Time	t	second	s
Torque	T	newton metre	N m
Velocity	v	metre per second	m/s
angular	ω	radian per second	rad/s
Voltage	V	volt	V
Volume	V	cubic metre	m^3
Volume, rate of flow	V	cubic metre per second	m^3/s
Wavelength	λ	metre	m
Work	W	joule	J
Young's modulus of elasticity	E	Pascal	Pa

7 Basic d.c. circuit theory

Standard symbols for electrical components

1 Symbols are used for components in electrical circuit diagrams and some of the more common ones are shown in *Figure 7.1*.

2   (i) All substances are made from **elements** and the smallest particle to which an element can be reduced is called an **atom**.

(ii) An atom consists of **electrons** which can be considered to be orbiting around a central **nucleus** containing **protons** and **neutrons**.

(iii) An electron possesses a **negative charge**, a proton a **positive charge** and a neutron has no charge.

(iv) There is a force of **attraction** between oppositely charged bodies and a force of **repulsion** between similarly charged bodies.

(v) The **force** between two charged bodies depends on the amount of charge on the bodies and their distance apart.

(vi) **Conductors** are materials having electrons that are loosely connected to the nucleus and can easily move through the material from one atom to another. **Insulators** are materials whose electrons are held firmly to their nucleus.

(vii) A drift of electrons in the same direction constitutes an **electric current**.

(viii) The unit of charge is the **coulomb, C**, and when 1 coulomb of charge is transferred in 1 second a current of 1 ampere flows in the conductor. This electric current I is the rate of flow of charge in a circuit. The unit of current is the **ampere, A**.

(ix) For a continuous current to flow between two points in a circuit a **potential difference (p.d.)** or voltage, **V**, is required between them; a complete conducting path is necessary to and from the source of electrical energy. The unit of p.d. is the **volt, V**.

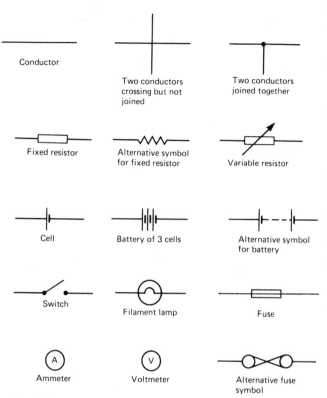

Figure 7.1

(x) *Figure 7.2* shows a cell connected across a filament lamp. Current flow, by convention, is considered as flowing from the positive terminal of the cell, around the circuit to the negative terminal.

3   The flow of electric current is subject to friction. This friction, or opposition, is called **resistance** R and is the property of a conductor that limits current. The unit of resistance is the ohm, Ω. 1 ohm is defined as the resistance which will have a current of 1 ampere flowing through it when 1 volt is connected across it,

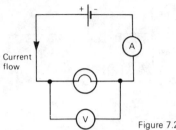

Figure 7.2

i.e. **resistance** $R = \dfrac{\textbf{potential difference}}{\textbf{current}}$, i.e. $R = \dfrac{V}{I}$

4   The reciprocal of resistance is called **conductance** and is measured in siemens (S). Thus

conductance in siemens, $G = \dfrac{1}{R}$

where R is the resistance in ohms.

Electrical measuring instruments

5   (i) An **ammeter** is an instrument used to measure current and must be connected **in series** with the circuit. *Figure 7.2* shows an ammeter connected in series with the lamp to measure the current flowing through it. Since all the current in the circuit passes through the ammeter it must have a very **low resistance**.

(ii) A **voltmeter** is an instrument used to measure p.d. and must be connected **in parallel** with the part of the circuit whose p.d. is required. In *Figure 7.2*, a voltmeter is connected in parallel with the lamp to measure the p.d. across it. To avoid a significant current flowing through it a voltmeter must have a very **high resistance**.

(iii) An **ohmmeter** is an instrument for measuring resistance.

(iv) A **multimeter**, or universal instrument, may be used to measure voltage, current and resistance. An 'Avometer' is a typical example.

(v) The **cathode ray oscilloscope (CRO)** may be used to observe waveforms and to measure voltages and currents. The display of a CRO involves a spot of light moving across a screen. The amount by which the spot is deflected from its initial position depends on the p.d. applied to the terminals of the CRO and the range selected. The displacement is calibrated in 'volts per cm'. For example, if the spot is deflected 3 cm and the volts/cm switch is on 10 V/cm then the magnetude of the p.d. is 3 cm × 10 V/cm, i.e., 30 V. (See chapter 24, page 190).

Linear and non-linear devices

6   *Figure 7.3* shows a circuit in which current I can be varied by the variable resistor R_2. For various settings of R_2, the current flowing in resistor R_1, displayed on the ammeter, and the p.d.

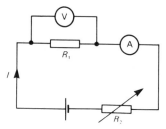

Figure 7.3

across R_1, displayed on the voltmeter, are noted and a graph is plotted of p.d. against current. The result is shown in *Figure 7.4(a)* where the straight line graph passing through the origin indicates that current is directly proportional to the p.d. Since the gradient i.e. p.d./current is constant, resistance R_1 is constant. A resistor is thus an example of a **linear device**.

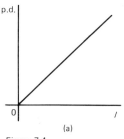

(a)

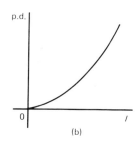

(b)

Figure 7.4

If the resistor R_1 in *Figure 7.3* is replaced by a component such as a lamp then the graph shown in *Figure 7.4(b)* results when values of p.d. are noted for various current readings. Since the gradient is changing the lamp is an example of a **non-linear device**.

7   **Ohm's law** states that the current I flowing in a circuit is directly proportional to the applied voltage V and inversely proportional to the resistant R, provided the resistance remains constant. Thus:

$$\boxed{I = \frac{V}{R}} \quad \text{or} \quad \boxed{V = IR} \quad \text{or} \quad \boxed{R = \frac{V}{I}}$$

8   (i) A **conductor** is a material having a low resistance which allows electric current to flow in it. All metals are conductors and some examples include copper, aluminium, brass, platinum, silver, gold and also carbon.

(ii) An **insulator** is a material having a high resistance which does not allow electric current to flow in it. Some examples of insulators include plastic, rubber, glass, porcelain, air, paper, cork, mica, ceramics and certain oils.

Series circuit

9   *Figure 7.5* shows three resistors R_1, R_2 and R_3 connected end to end, i.e., in series with a battery source of V volts. Since the

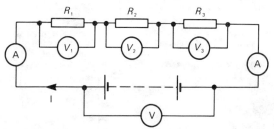

Figure 7.5

circuit is closed a current I will flow and the p.d. across each resistor may be determined from the voltmeter readings V_1, V_2 and V_3. In a series circuit:

24

(a) the current I is the same in all parts of the circuit and hence the same reading is found on each of the ammeters shown, and

(b) the sum of the voltages V_1, V_2 and V_3 is equal to the total applied voltage, V, i.e.

$$V = V_1 + V_2 + V_3.$$

From Ohm's law:

$$V_1 + IR_1, \quad V_2 = IR_2, \quad V_3 = IR_3 \text{ and } V = IR$$

where R is the total circuit resistance.

Since $V = V_1 + V_2 + V_3$

then $IR = IR_1 + IR_2 + IR_3.$

Dividing throughout by I gives

$$R = R_1 + R_2 + R_3.$$

Thus for a series circuit, the total resistance is obtained by adding together the values of the separate resistances.

10   The voltage distribution for the circuit shown in *Figure 7.6(a)* is given by:

$$V_1 = \left(\frac{R_1}{R_1 + R_2} \right) V$$

$$V_2 = \left(\frac{R_2}{R_1 + R_2} \right) V$$

The circuit shown in *Figure 7.6(b)* is often referred to as a **potential divider** circuit. Such a circuit can consist of a number of similar elements in series connected across a voltage source, voltages being taken from connections in-between the elements. Frequently the potential divider consists of two resistors as shown in *Figure 7.6(b)* where

$$V_{OUT} = \left(\frac{R_2}{R_1 + R_2} \right) V_{IN}$$

Where a continuously variable voltage is required from a fixed supply a single resistor with a sliding contact is used. Such a device is known as a **potentiometer**.

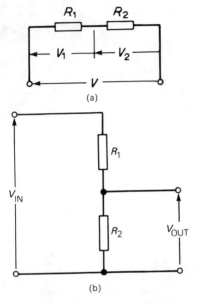

Figure 7.6

Parallel circuit

11   *Figure 7.7* shows three resistors, R_1, R_2 and R_3 connected across each other, i.e. in parallel, across a battery source of V volts. In a parallel circuit:

    (a) the sum of the currents I_1, I_2 and I_3 is equal to the total circuit current, I, i.e. $I = I_1 + I_2 + I_3$, and

    (b) the source p.d., V volts, is the same across each of the resistors.

From Ohm's law:

$$I_1 = \frac{V}{R_1},\ I_2 = \frac{V}{R_2},\ I_3 = \frac{V}{R_3}\ \text{and}\ I = \frac{V}{R}$$

where R is the total circuit resistance.
Since $I = I_1 + I_2 + I_3$

Then $\dfrac{V}{R} = \dfrac{V}{R_1} + \dfrac{V}{R_2} + \dfrac{V}{R_3}$

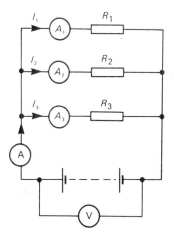

Figure 7.7

Dividing throughout by V gives:

$$\frac{1}{R} = \frac{1}{R_1} + \frac{1}{R_2} + \frac{1}{R_3}$$

This equation must be used when finding the total resistance R of a parallel circuit.

12 For the special case of two resistors in parallel:

$$\frac{1}{R} = \frac{1}{R_1} + \frac{1}{R_2} = \frac{R_2 + R_1}{R_1 R_2}$$

Hence

$$R = \frac{R_1 R_2}{R_1 + R_2} \left(\text{i.e. } \frac{\text{product}}{\text{sum}} \right)$$

13 The current division for the circuit shown in *Figure 7.8* is given by:

$$I_1 = \left(\frac{R_2}{R_1 + R_2} \right) I$$

$$I_2 = \left(\frac{R_1}{R_1 + R_2} \right) I$$

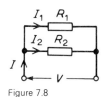

Figure 7.8

27

Wiring lamps in series and in parallel

Series connection

14   *Figure 7.9* shows three lamps, each rated at 240 V, connected in series across a 240 V supply.

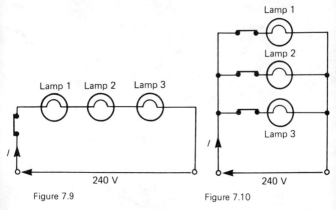

Figure 7.9   Figure 7.10

(i) Each lamp has only $\frac{240}{3}$ V, i.e. 80 V across it and thus each lamp glows dimly.
(ii) If another lamp of similar rating is added in series with the other three lamps then each lamp now has $\frac{240}{4}$ V, i.e. 60 V across it and each now glows even more dimly.
(iii) If a lamp is removed from the circuit or if a lamp develops a fault (i.e. an open circuit) or if the switch is opened then the circuit is broken, no current flows, and the remaining lamps will not light up.
(iv) Less cable is required for a series connection than for a parallel one.

The series connection of lamps is usually limited to decorative lighting such as for Christmas tree lights.

Parallel connection

Figure 7.10 shows three similar lamps, each rated at 240 V, connected in parallel across a 240 V supply.

(i) Each lamp has 240 V across it and thus each will glow brilliantly at their rated voltage.

(ii) If any lamp is removed from the circuit or develops a fault (open circuit) or a switch is opened, the remaining lamps are unaffected.

(iii) The addition of further similar lamps in parallel does not affect the brightness of the other lamps.

(iv) More cable is required for parallel connection than for a series one.

The parallel connection of lamps is the most widely used in electrical installations.

15 **Power** P in an electrical circuit is given by the product of potential difference V and current I. The unit of power is the **watt, W**.

$$\text{Hence } \boldsymbol{P = V \times I} \quad \textbf{watts.} \tag{1}$$

From Ohm's law, $V = IR$

Substituting for V in (1) gives:

$$P = (IR) \times I \text{ i.e.}$$

i.e. $\boldsymbol{P = I^2 R}$ **watts.**

Also, from Ohms law, $I = \dfrac{V}{R}$

Substituting for I in (1) gives:

$$P = V \times \left(\frac{V}{R}\right) \text{ i.e. } \boldsymbol{P = \frac{V^2}{R}} \textbf{ watts.}$$

There are thus three possible formulae which may be used for calculating power.

16 **Electrical energy = power × time.**

If the power is measured in watts and the time in seconds then the unit of energy is watt-seconds or **joules**. If the power is measured in kilowatts and the time in hours then the unit of energy is kilowatt-hours, often called the 'unit of electricity'. The 'electricity meter' in the home records the number of kilowatt-hours used and is thus an energy meter.

$$\begin{aligned} 1 \text{ kWh} &= 1000 \text{ watt hours} \\ &= 1000 \times 3600 \text{ watt seconds or joules} \\ &= 3\,600\,000 \text{ J.} \end{aligned}$$

17   (i) The **three main effects of an electric current** are:

(a) magnetic effect;

(b) chemical effect;

(c) heating effect.

(ii) Some practical applications of the effects of an electric current include:

Magnetic effect: bells, relays, motors, generators, transformers, telephones, car-ignition and lifting magnets.

Chemical effect: primary and secondary cells and electroplating.

Heating effect: cookers, water heaters, electric fires, irons, furnaces, kettles and soldering irons.

18 A **fuse** is used to prevent overloading of electrical circuits.

The fuse, which is made of material having a low melting point, utilizes the heating effect of an electric current. A fuse is placed in an electrical circuit and if the current becomes too large the fuse wire melts and so breaks the circuit. A circuit diagram symbol for a fuse is shown in *Figure 7.1*, page 21.

Resistance variation

19 The resistance of an electrical conductor depends on four factors, these being:

(a) the length of the conductor;
(b) the cross-sectional area of the conductor;
(c) the type of material; and
(d) the temperature of the material.

20   (i) Resistance, R, is directly proportional to length, l, of a conductor, i.e. $R \propto l$. Thus, for example, if the length of a piece of wire is doubled, then the resistance is doubled.

(ii) Resistance, R, is inversely proportional to cross-sectional area, a, of conductor, i.e. $R \propto (1/a)$. Thus, for example, if the cross-sectional area of a piece of wire is doubled then the resistance is halved.

(iii) Since $R \propto l$ and $R \propto (1/a)$ then $R \propto (1/a)$. By inserting a constant of proportionality into this relationship the type of material used may be taken into account. The constant of proportionality is known as the **resistivity** of the material and is given the symbol ρ (rho).

Thus, resistance, $$\boxed{R = \frac{\rho l}{a} \text{ ohms}}$$

ρ is measured in ohm metres (Ωm). The value of the resistivity is that resistance of a unit cube of the material measured between opposite faces of the cube.

(iv) Resistivity varies with temperature and some typical values of resistivities measured at about room temperature are given below:

Copper, 1.7×10^{-8} Ωm (or 0.017 $\mu\Omega$m)
Aluminium 2.6×10^{-8} Ωm (or 0.026 $\mu\Omega$m)
Carbon (graphite) 10×10^{-8} Ωm (0.10 $\mu\Omega$m)
Glass 1×10^{10} Ωm (or 10^4 $\mu\Omega$m)
Mica 1×10^{13} Ωm (or 10^7 $\mu\Omega$m).

Note that good conductors of electricity have a low value of resistivity and good insulators have a high value of resistivity.

21

(i) In general, as the temperature of a material increases, most conductors increase in resistance, insulators decrease in resistance whilst the resistance of some special alloys remain almost constant.

(ii) The **temperature coefficient of resistance** of a material is the increase in the resistance of a 1 Ω resistor of that material when it is subjected to a rise of temperature of 1°C. The symbol used for the temperature coefficient of resistance is α (alpha). Thus, if some copper wire of resistance 1 Ω is heated through 1°C and its resistance is then measured as 1.0043 Ω then

$\alpha = 0.0043$ Ω/Ω°C for copper.

The units are usually expressed only as 'per °C', i.e. $\alpha = 0.0043/$°C for copper. If the 1 Ω resistor of copper is heated through 100°C then the resistance at 100°C would be

$1 + 100 \times 0.0043 = 1.43$ Ω.

(iii) If the resistance of a material at 0°C is known, the resistance at any other temperature can be determined from:

$$\boxed{R_\theta = R_0(1 + \alpha_0\theta)}$$

where R_0 = resistance at 0°C

R_θ = resistance at temperature θ°C;

α_0 = temperature coefficient of resistance at 0°C.

(iv) If the resistance at 0°C is not known, but is known at some other temperature θ_1, then the resistance at any temperature can be found as follows:

$R_1 = R_0(1 + \alpha_0\theta_1)$ and $R_2 = R_0(1 + \alpha_0\theta_2)$

Dividing one equation by the other gives:

$$\boxed{\frac{R_1}{R_2} = \frac{1 + \alpha_0\theta_1}{1 + \alpha_0\theta_2}}$$

where R_2 = resistance at temperature θ_2.

(v) If the resistance of a material at room temperature (approximately 20°C), R_{20}, and the temperature co-efficient of resistance at 20°C, α_{20} are known then the resistance R_θ at temperature θ°C is given by:

$$R_\theta = R_{20}[1 + \alpha_{20}(\theta - 20)]$$

(vi) Some typical values of temperature coefficient of resistance measured at 0°C are given below:

Copper	0.0043/°C
Aluminium	0.0038/°C
Nickel	0.0062/°C
Carbon	−0.00048/°C
Constantan	0
Eureka	0.00001/°C

(Note that the negative sign for carbon indicates that its resistance falls with increase of temperature.)

8 d.c. circuit analysis

1   The laws which determine the currents and voltage drops in
d.c. networks are

(a) Ohm's law;
(b) the laws for resistors in series and in parallel; and
(c) Kirchhoff's laws.

Kirchhoff's laws

2   Kirchhoff's laws state:

(a) Current law

*At any junction in an electric circuit the total current flowing
towards that junction is equal to the total current flowing away
from the junction, i.e.* $\Sigma I = 0$

Thus, referring to *Figure 8.1*:

$$I_1 + I_2 = I_3 + I_4 + I_5$$
$$\text{or} \quad I_1 + I_2 - I_3 - I_4 - I_5 = 0$$

(b) Voltage law

*In any closed loop in a network, the algebraic sum of the voltage
drops (i.e. products of current and resistance) taken around the
loop is equal to the resultant emf acting in that loop.*

Thus referring to *Figure 8.2*:

$$E_1 - E_2 = IR_1 + IR_2 + IR_3$$

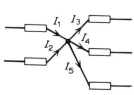

Figure 8.1

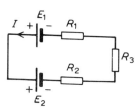

Figure 8.2

(Note that if current flows away from the positive terminal of a source, that source is considered by convention to be positive. Thus moving anticlockwise around the loop of *Figure 8.2*, E_1 is positive and E_2 is negative.)

For example, using Kirchhoff's laws to determine the current flowing in each branch of the network shown in *Figure 8.3*, the procedure is as follows:

(i) Use Kirchhoff's current law and label current directions on the original circuit diagram. The directions chosen are arbitrary, but it is usual, as a starting point, to assure that current flows from the positive terminals of the batteries. This is shown in *Figure 8.4* where the three branch currents are expressed in terms of I_1 and I_2 only, since the current through R is $I_1 + I_2$.

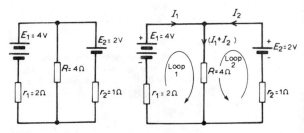

Figure 8.3   Figure 8.4

(ii) Divide the circuit into two loops and apply Kirchhoff's voltage law to each. From loop 1 of *Figure 8.4*, and moving in a clockwise direction as indicated (the direction chosen does not matter) gives

$E_1 = I_1 r_1 + (I_1 + I_2)R,$
i.e. $4 = 2I_1 + 4(I_1 + I_2),$
i.e. $6I_1 + 4I_2 = 4$ (1)

From loop 2 of *Figure 8.4*, and moving in an anticlockwise direction as indicated (once again, the choice of direction does not matter; it does not have to be in the same direction as that chosen from the first loop), gives:

$E_2 = I_2 r_2 + (I_1 + I_2)R,$
i.e. $2 = I_2 + 4(I_1 + I_2),$
i.e. $4I_1 + 5I_2 = 2$ (2)

(iii) Solve equations (1) and (2) for I_1 and I_2.

$2 \times (1)$ gives: $12I_1 + 8I_2 = 8$ (3)

$3 \times (2)$ gives: $12I_1 + 15I_2 = 6$ (4)

$(3) - (4)$ gives: $-7I_2 = 2$ and $I_2 = -\dfrac{2}{7} = $ **−0.286 A**

(i.e. I_2 is flowing in the opposite direction to that shown in *Figure 8.4*).

From (1) $6I_1 + 4(-0.286) = 4$

          $6I_2 = 4 + 1.144$

Hence $I_1 = \dfrac{5.144}{6} = $ **0.857 A**

Current flowing through $R = I_1 + I_2 = 0.857 + (-0.286)$
                               $= 0.571$ A.

General hints on simple d.c. circuit analysis

3   (i) The open-circuit voltage, E, across terminals AB in *Figure 8.5* is equal to 10 V, since no current flows through the 2 Ω resistor and hence no voltage drop occurs.

(ii) The open-circuit voltage, E, across terminals AB in *Figure 8.6(a)* is the same as the voltage across the 6 Ω resistor. The circuit may be redrawn as shown in *Figure 8.6(b)*.

$E = \left(\dfrac{6}{6+4}\right)(50)$

by voltage division
in a series circuit.

i.e. $E = 30$ V

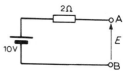

Figure 8.5

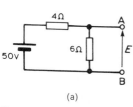

(a)

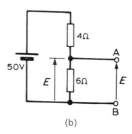

(b)

Figure 8.6

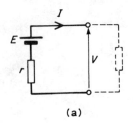 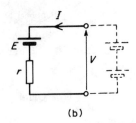

Figure 8.7

(iii) For the circuit shown in *Figure 8.7(a)* representing a practical source supplying energy, $V = E - Ir$ where E is the battery emf, V is the battery terminal voltage and r is the internal resistance of the battery. For the circuit shown in *Figure 8.7(b)*,

$$V = E - (-I)r, \text{ i.e. } V = E + Ir$$

(iv) The resistance 'looking-in' at terminals AB in *Figure 8(a)* is obtained by reducing the circuit in stages as shown in *Figures 8.8(b) to (d)*. Hence the equivalent resistance across AB is 7Ω.

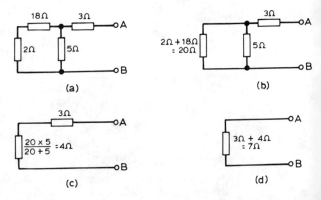

Figure 8.8

36

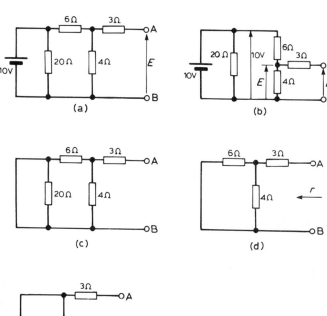

Figure 8.9

(v) For the circuit shown in *Figure 8.9(a)*, the 3 Ω resistor carries no current and the p.d. across the 20Ω resistor is 10 V. Redrawing the circuit gives *Figure 8.9(b)*, from which

$$E = \left(\frac{4}{4+6}\right) \times 10 = 4 \text{ V}.$$

(vi) If the 10 V battery in *Figure 8.9(a)* is removed and replaced by a short-circuit, as shown in *Figure 8.9(c)*, then the 20 Ω resistor may be removed. The reason for this is that a short-circuit has zero resistance, and 20 Ω in parallel with zero ohms gives an equivalent resistance of

$$\frac{20 \times 0}{20 + 0}, \text{ i.e. } 0 \ \Omega.$$

The circuit is then as shown in *Figure 8.9(d)*, which is redrawn in *Figure 8.9(e)*. From *Figure 8.9(e)*, the equivalent resistance across AB,

$$r = \frac{6 \times 4}{6 + 4} + 3 = 2.4 + 3 = 5.4 \ \Omega$$

(vii) To find the voltage across AB in *Figure 8.10*: Since the 20 V supply is across the 5Ω and 15 Ω resistors in series then, by voltage division, the voltage drop across AC,

$$V_{AC} = \left(\frac{5}{5 + 15}\right)(20) = 5 \text{ V}.$$

Similarly,

$$V_{CB} = \left(\frac{12}{12 + 3}\right)(20) = 16 \text{ V}$$

V_C is at a potential of $+20$ V.

$$V_A = V_C - V_{AC} = +20 - 5$$
$$= 15 \text{ V and}$$
$$V_B = V_C - V_{BC}$$
$$= +20 - 16 = 4 \text{ V}.$$

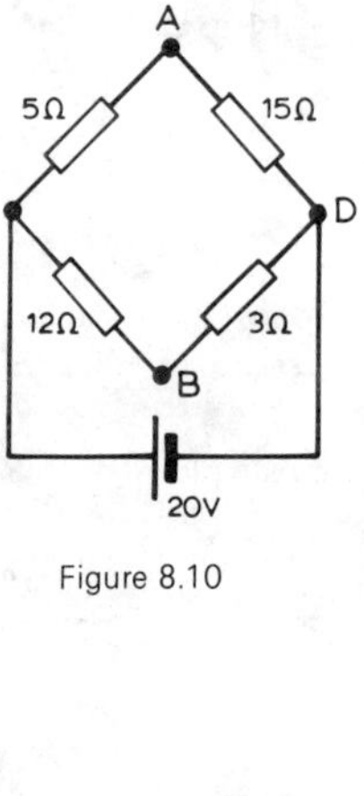

Figure 8.10

Hence the voltage between AB is

$$V_A - V_B = 15 - 4 = 11 \text{ V}$$

and current would flow from A to B since A has a higher potential than B.

(viii) In *Figure 8.11(a)*, to find the equivalent resistance across AB, the circuit may be redrawn as in *Figures 8.11(b)* and *(c)*. From *Figure 8.11(c)*, the equivalent resistance across AB

$$= \frac{5 \times 15}{5 + 15} + \frac{12 \times 3}{12 + 3} = 3.75 + 2.4 = 6.15 \ \Omega$$

4   There are a number of circuit theorems which have been developed for solving problems in d.c. electrical networks.

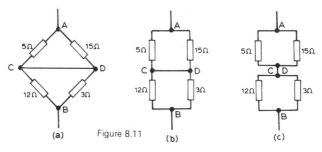

Figure 8.11

These include:

       (i) the superposition theorem;
      (ii) Thévénins theorem;
     (iii) Norton's theorem, and
     (iv) the maximum power transfer theorem.

5   **The superposition theorem** states:

    *'In any network made up of linear resistances and containing more than
    one source of emf, the resultant current flowing in any branch is the
    algebraic sum of the currents that would flow in that branch if each
    source was considered separately, all other sources being replaced at that
    time by their respective internal resistances.'*

For example, to determine the current in
each branch of the network shown in
Figure 8.12 using the superposition theorem, the procedure is as follows.

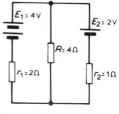

Figure 8.12

       (i) Redraw the original circuit
       with source E_2 removed, being
       replaced by r_2 only, as shown in
       Figure 8.13(a).
      (ii) Label the currents in each
      branch and their directions as
      shown in *Figure 8.13(a)* and
determine their values. (Note that the choice of current
directions depends on the battery polarity, which, by
convention is taken as flowing from the positive battery
terminal as shown.) R in parallel with r_2 gives an equivalent resistance of

$$\frac{4 \times 1}{4 + 1} = 0.8 \ \Omega.$$

39

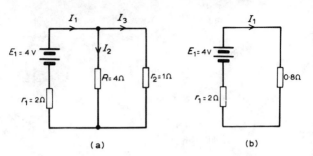

Figure 8.13

From the equivalent circuit of *Figure 8.13(b)*.

$$I_1 = \frac{E_1}{r_1 + 0.8} = \frac{4}{2 + 0.8} = 1.429 \text{ A}$$

From *Figure 8.13(a)*,

$$I_2 = \left(\frac{1}{4+1}\right)I_1 = \frac{1}{5}(1.429) = 0.286 \text{ A and}$$

$$I_3 = \left(\frac{4}{4+1}\right)I_1 = \frac{4}{5}(1.429) = 1.143 \text{ A}$$

(iii) Redraw the original circuit with source E_1 removed, being replaced by r_1 only, as shown in *Figure 8.14(a)*.
(iv) Label the currents in each branch and their directions as shown in *Figure 8.14(a)* and determine their

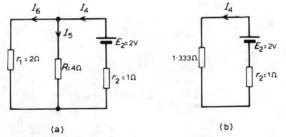

(a)

(b)

Figure 8.14

values. r_1 in parallel with R gives an equivalent resistance of

$$\frac{2 \times 4}{2+4} = \frac{8}{6} = 1.333 \ \Omega$$

From the equivalent circuit of *Figure 8.14(b)* :

$$I_4 = \frac{E_2}{1.333 + r_2} = \frac{2}{1.333 + 1}$$
$$= 0.857 \text{ A}$$

From *Figure 8.14(a)*

$$I_5 = \left(\frac{2}{2+4}\right)I_4 = \frac{2}{6}(0.857) = 0.286 \text{ A}$$

$$I_6 = \left(\frac{4}{2+4}\right)I_4 = \frac{4}{6}(0.857) = 0.571 \text{ A}$$

(v) Superimpose *Figure 8.14(a)* onto *Figure 8.13(a)* as shown in *Figure 8.15(a)*.

(vi) Determine the algebraic sum of the currents flowing in each branch. Resultant current flowing through source 1, i.e.

$$I_1 - I_6 = 1.429 - 0.571$$

$$= \textbf{0.858 A (discharging)}$$

Resultant current flowing through source 2, i.e.

$$I_4 - I_3 = 0.857 - 1.143$$

$$= \textbf{-0.286 A (charging)}$$

Resultant current flowing through resistor R, i.e.

$$I_2 + I_5 = 0.286 + 0.286 = \textbf{0.572 A}.$$

The resultant currents with their directions are shown in *Figure 8.16*.

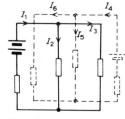

Figure 8.15

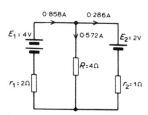

Figure 8.16

6   (a) **Thévénin's theorem** states:

> '*The current in any branch of a network is that which would result if an emf, equal to the p.d. across a break made in the branch, were introduced into the branch, all other emf's being removed and represented by the internal resistances of the sources.*'

(b) The procedure adopted when using Thévénin's theorem is summarised below. To determine the current in any branch of an active network (i.e. one containing a source of emf):

(i) remove the resistance R from that branch,

(ii) determine the open-circuit voltage, E, across the break,

(iii) remove each source of emf and replace them by their internal resistances and then determine the resistance, r, 'looking-in' at the break.

(iv) determine the value of the current from the equivalent circuit shown in *Figure 8.17*,

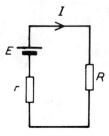

Figure 8.17

i.e. $I = \dfrac{E}{R+r}$

For example, using Thévénin's theorem to determine the current flowing in the 4Ω resistor shown in *Figure 8.18(a)*, using the above procedure:

(i) The $4\ \Omega$ resistor is removed from the circuit as shown in *Figure 8.18(b)*.

(ii) Current $I_1 = \dfrac{E_1 - E_2}{r_1 + r_2} = \dfrac{4-2}{2+1} = \dfrac{2}{3}$ A.

P.d. across AB, $E = E_1 - I_1 r_1 = 4 - \dfrac{2}{3}(2) = 2\dfrac{2}{3}$ V

$\left(\text{Alternatively, p.d. across AB, } E = E_2 - I_1 r_2 = 2 - \left(-\dfrac{2}{3}\right)(1) \right.$
$\left. \hspace{5.5cm} = 2\dfrac{2}{3}\ \text{V} \right).$

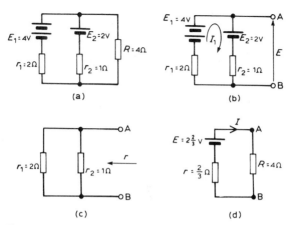

Figure 8.18

(iii) Removing the sources of emf gives the circuit shown in *Figure 8.18(c)*, from which resistance $r = \dfrac{2 \times 1}{2 + 1} = \dfrac{2}{3}\ \Omega$.

(iv) The equivalent Thévenin's circuit is shown in *Figure 8.18(d)* from which current

$$I = \frac{E}{r + R} = \frac{2\frac{2}{3}}{\frac{2}{3} + 4} = \frac{8/3}{14/3} = \frac{8}{14} = 0.571\ \text{A}.$$

(c) Thévenin's theorem can be used to analyse part of a circuit, and in complicated networks the principle of replacing the supply by a constant voltage source in series with a resistance is very useful.

7   (a) **Norton's theorem** states:

> '*The current that flows in any branch of a network is the same as that which would flow in the branch if it was connected across a source of electricity, the short-circuit current of which is equal to the current that would flow in a short-circuit across the branch, and the internal resistance of which is equal to the resistance which appears across the open-circuited branch terminals.*'

(b) The procedure adopted when using Norton's theorem is summarised below. To determine the current in any branch AB of an active network:

(i) short-circuit that branch,

(ii) determine the short-circuit current, I_{sc},

(iii) remove each source of emf and replace them by their internal resistances (or, if a current source exists replace with an open circuit), then determine the resistance, R, "looking-in" at a break made between A and B,

(iv) determine the value of the current from the equivalent circuit shown in *Figure 8.19*,

i.e. $I = \left(\dfrac{R}{R + R_L}\right) I_{sc}$.

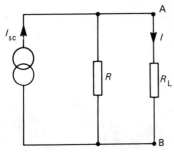

Figure 8.19

For example, to determine the current flowing in the 4 Ω resistor of *Figure 8.18(a)* using Norton's theorem by the above procedure:

(i) the branch containing the 4 Ω resistor is short-circuited as shown in *Figure 8.20*.

(ii) the short-circuit current I_{sc} is given by:

$$I_{sc} = I_1 + I_2$$
$$= \frac{4}{2} + \frac{2}{1} = 2 + 2 = 4 \text{ A}.$$

(iii) resistance $R = \frac{2}{3}$ Ω (same as procedure (iii) of para. 6)

(iv) from the equivalent Norton circuit shown in *Figure 8.21*

current $I = \left(\dfrac{R}{R + R_L}\right) I_{sc} = \left(\dfrac{\frac{2}{3}}{\frac{2}{3} + 4}\right)(4) = \textbf{0.571 A}.$

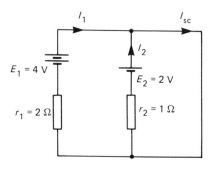

Figure 8.20

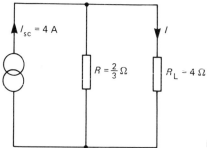

Figure 8.21

8   A Thévénin equivalent circuit having emf E and internal resistance r can be replaced by a Norton equivalent circuit containing a current generator I_{sc} and internal resistance R, where:

$$R = r, \quad E = I_{sc}R \quad \text{and} \quad I_{sc} = \frac{E}{r}.$$

Thus,

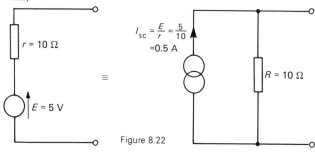

Figure 8.22

45

and

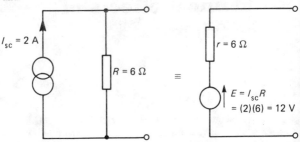

Figure 8.23

9 (a) **The maximum power transfer theorem** states:

'*The power transferred from a supply source to a load is at its maximum when the resistance of the load is equal to the internal resistance of the source.*'

Hence, in *Figure 8.24*, when $R = r$ the power transferred from the source to the load is a maximum.

(b) Varying a load resistance to be equal, or almost equal, to the source internal resistance is called **resistance matching**. Examples where resistance matching is important include coupling an aerial to a transmitter or receiver, or in coupling a loudspeaker to an amplifier where coupling transformers may be used to give maximum power transfer (see chapter 21, para 11, page 161).

With d.c. generators or secondary cells, the internal resistance is usually very small. In such cases, if an attempt is made to make the load resistance as small as the source internal resistance, overloading of the source results.

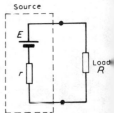

Figure 8.24

9 Chemical effects of electricity

1   A material must contain **charged particles** to be able to conduct electric current. In solids, the current is carried by **electrons**. Copper, lead, aluminium, iron and carbon are some examples of solid conductors. In **liquids and gases**, the current is carried by the part of a molecule which has acquired an electric charge, called **ions**. These can possess a positive or negative charge, and examples including hydrogen ion $H^+ \ldots$, copper ion $Cu^{++} \ldots$ and hydroxyl ion $OH^- \ldots$ Distilled water contains no ions and is a poor conductor of electricity whereas salt water contains ions and is a fairly good conductor of electricity.

2   (i) **Electrolysis** is the decomposition of a liquid compound by the passage of electric current through it. Practical applications of electrolysis include the electroplating of metals (see para. 3), the refining of copper and the extraction of aluminium from its ore.

(ii) An **electrolyte** is a compound which will undergo electrolysis. Examples include salt water, copper sulphate and sulphuric acid.

(iii) The **electrodes** are the two conductors carrying current to the electrolyte. The positive-connected electrode is called the **anode** and the negative-connected electrode the **cathode**.

(iv) When two copper wires connected to a battery are placed in a beaker containing a salt water solution, then current will flow through the solution. Bubbles appear around the wires as the water is changed into hydrogen and oxygen by electrolysis.

3   **Electroplating** uses the principle of electrolysis to apply a thin coat of one metal to another metal. Some practical applications include the tin-plating of steel, silver-plating of nickel alloys and chromium-plating of steel. If two copper electrodes connected to a battery are placed in a beaker containing copper sulphate as the electrolyte it is found that the cathode (i.e. the electrode connected to the negative terminal of the battery) gains copper whilst the anode loses copper.

4   The purpose of an **electric cell** is to convert chemical energy into electrical energy. A **simple cell** comprises two dissimilar conductors (electrodes) in an electrolyte. Such a cell is shown in *Figure 9.1*, comprising copper and zinc electrodes. An electric

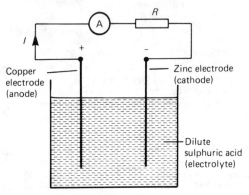

Figure 9.1

current is found to flow between the electrodes. Other possible electrode pairs exist, including zinc-lead and zinc-iron. The electrode potential (i.e. the p.d. measured between the electrodes) varies for each pair of metals.

By knowing the emf of each metal with respect to some standard electrode the emf of any pair of metals may be determined. The standard used is the hydrogen electrode. The **electrochemical series** is a way of listing elements in order of electrical potential, and *Table 9.1* shows a number of elements in such a series.

Table 9.1   Part of the electrochemical series

Potassium
sodium
aluminium
zinc
iron
lead
hydrogen
copper
silver
carbon

In a simple cell two faults exist – those due to polarization and local action.

(a) If the simple cell shown in *Figure 9.1* is left connected for some time, the current I decreases fairly rapidly. This is because of the formation of a film of hydrogen bubbles on the copper anode. This effect is known as the polarization of the cell. The hydrogen prevents full contact between the copper electrode and the electrolyte and this increases the internal resistance of the cell. The effect can be overcome by using a chemical depolarizing agent or depolarizer, such as potassium dichromate which removes the hydrogen bubbles as they form. This allows the cell to deliver a steady current.

(b) When commercial zinc is plated in dilute sulphuric acid, hydrogen gas is liberated from it and the zinc dissolves. The reason for this is that impurities, such as traces of iron, are present in the zinc which set up small primary cells with the zinc. These small cells are short-circuited by the electrolyte, with the result that localised currents flow causing corrosion. This action is known as **local action** of the cell. This may be prevented by rubbing a small amount of mercury on the zinc surface, which forms a protection layer on the surface of the electrode.

5 When two metals are used in a simple cell the electrochemical series may be used to predict the behaviour of the cell:

(i) The metal that is higher in the series acts as the negative electrode, and vice-versa. For example, the zinc electrode in the cell shown in *Figure 9.1* is negative and the copper electrode is positive.

(ii) The greater the separation in the series between the two metals the greater is the emf produced by the cell.

6 The electrochemical series is representative of the order of reactivity of the metals and their compounds.

(i) The higher metals in the series react more readily with oxygen and vice-versa.

(ii) When two metal electrodes are used in a simple cell the one that is higher in the series tends to dissolve in the electrolyte.

7 (i) **Corrosion** is the gradual destruction of a metal in a damp atmosphere by means of simple cell action. In addition to the presence of moisture and air required for rusting, an electrolyte, an anode and a cathode are required for corrosion. Thus, if metals widely spaced in the

49

electrochemical series, are used in contact with each other in the presence of an electrolyte, corrosion will occur. For example, if a brass valve is fitted to a heating system made of steel, corrosion will occur.

(ii) The **effects of corrosion** include the weakening of structures, the reduction of the life of components and materials, the wastage of materials and the expense of replacement.

(iii) Corrosion may be **prevented** by coating with paint, grease, plastic coatings and enamels, or by plating with tin or chromium. Also, iron may be galvanised, i.e. plated with zinc, the layer of zinc helping to prevent the iron from corroding.

8 (i) The **electromotive force (emf)**, E, of a cell is the p.d. between its terminals when it is not connected to a load (i.e. the cell is on 'no-load').

(ii) The emf of a cell is measured by using a **high resistance voltmeter** connected in parallel with the cell. The voltmeter must have a high resistance otherwise it will pass current and the cell will not be on no-load. For example, if the resistance of a cell is 1 Ω and that of a voltmeter 1 MΩ then the equivalent resistance of the circuit is 1 MΩ+1 Ω, i.e., approximately 1 MΩ, hence no current flows and the cell is not loaded.

(iii) The voltage available at the terminals of a cell falls when a load is connected. This is caused by the **internal resistance** of the cell which is the opposition of the material of the cell to the flow of current. The internal resistance acts in series with other resistances in the circuit. *Figure 9.2* shows a cell of emf E volts and internal resistance r, XY represents the terminals of the cell. When a load (shown as resistance R) is not connected, no current flows and the terminal p.d., $V = E$.

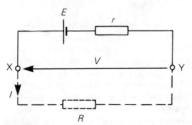

Figure 9.2

When R is connected a current I flows which causes a voltage drop in the cell, given by Ir. The p.d. available at the cell terminals is less than the emf of the cell and is given by: $V = E - Ir$. Thus if a battery of emf 12 V and internal resistance 0.01 Ω delivers a current of 100 A, the terminal p.d.,

$$V = 12 - (100)(0.01) = 12 - 1 = 11 \text{ V}.$$

(iv) When a current is flowing in the direction shown in *Figure 9.2* the cell is said to be discharging $(E > V)$.

(v) When a current flows in the opposite direction to that shown in *Figure 9.2* the cell is said to be charging $(V > E)$.

9   A **battery** is a combination of more than one cell. The cells in a battery may be connected in series or in parallel.

(i) **For cells connected in series:**
Total emf = sum of cell's emf's
Total internal resistance = sum of cell's internal resistances.

(ii) **For cells connected in parallel:**
If each cell has the same emf and internal resistance:
Total emf = emf of one cell.

Total internal resistance of n cells $= \dfrac{1}{n} \times$ internal resistance of one cell.

10   There are two main types of cell – primary cells and secondary cells.

(i) **Primary cells** cannot be recharged, that is, the conversion of chemical energy to electrical energy is irreversible and the cell cannot be used once the chemicals are exhausted. Examples of primary cells include the Lechlanché cell and the mercury cell.

A typical dry **Leclanché cell** is shown in *Figure 9.3.* Such a cell has an emf of about 1.5 V when new, but this falls rapidly if in continuous use due to polarization (see para. 4). The hydrogen film on the carbon electrode forms faster than can be dissipated by the depolarizer. The Lechlanché cell is suitable only for intermittent use, applications including torches, transistor radios, bells, indicator circuits, gas lighters, controlling switch-gear and so on. The cell is the most commonly used of primary cells, is cheap, requires little maintenance and has a shelf life of about two years.

A typical **mercury cell** is shown in *Figure 9.4.* Such a cell has an emf of about 1.3 V which remains constant

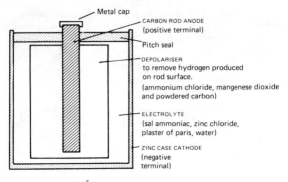

DRY LECLANCHÉ CELL

Figure 9.3

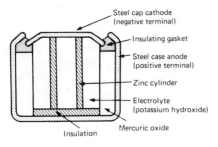

MERCURY CELL

Figure 9.4

for a relatively long time. Its main advantages over the Léclanche cell is its smaller size and its long shelf life.

Typical practical applications include hearing aids, medical electronics and for guided-missiles.

(ii) **Secondary cells** can be recharged after use, that is, the conversion of chemical energy to electrical energy is reversible and the cell may be used many times. Examples of secondary cells include the lead-acid cell and alkaline cells.

A typical **lead-acid cell** is constructed of:

    (i) A container made of glass, ebonite or plastic.
    (ii) Lead plates

(a) The negative plate (cathode) consists of spongy lead.

(b) The positive plate (anode) is formed by pressing lead peroxide into the lead grid.

(The plates are interleaved as shown in the plan view of *Figure 9.5* to increase their effective cross-sectional area and to minimise internal resistance.)

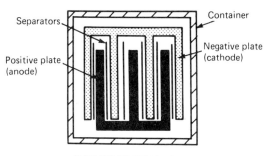

PLAN VIEW OF LEAD-ACID CELL

Figure 9.5

(iii) Separators made of glass, celluloid or wood.

(iv) An electrolyte which is a mixture of sulphur acid and distilled water.

The relative density (or specific gravity) of a lead-acid cell, which may be measured using a hydrometer, varies between about 1.26 when the cell is fully charged to about 1.19 when discharged. The terminal p.d. of a lead-acid cell is about 2 V.

When a cell supplies current to a load it is said to be discharging. During discharge:

(i) the lead peroxide (positive plate) and the spongy lead (negative plate) are converted into lead sulphate, and

(ii) the oxygen in the lead peroxide combines with hydrogen in the electrolyte to form water. The electrolyte is therefore weakened and the relative density falls.

The terminal p.d. of a lead-acid cell when fully discharged is about 1.8 V.

A cell is **charged** by connecting a d.c. supply to its terminals, the positive terminal of the cell being connected to the positive terminal of the supply. The charging current flows in the reverse direction to the discharge current and the chemical action is reversed. During charging:

> (i) the lead sulphate on the positive and negative plates is converted back to lead peroxide and lead respectively, and
>
> (ii) the water content of the electrolyte decreases as the oxygen released from the electrolyte combines with the lead of the positive plate. The relative density of the electrolyte thus increases.

The colour of the positive plate when fully charged is dark brown and when discharged is light brown. The colour of the negative plate when fully charged is grey and when discharged is light grey.

Practical applications of such cells include car batteries, telephone circuits and for traction purposes – such as milk delivery vans and fork lift trucks.

There are two main types of **alkaline cell** – the nickel-iron cell and the nickel-cadmium cell. In both types the positive plate is made of nickel hydroxide enclosed in finely perforated steel tubes, the resistance being reduced by the addition of pure nickel or graphite. The tubes are assembled into nickel-steel plates.

In the **nickel-iron cell**, (sometimes called the **Edison cell** or **nife cell**), the negative plate is made of iron oxide, with the resistance being reduced by a little mercuric oxide, the whole being enclosed in perforated steel tubes and assembled in steel plates. In the nickel-cadmium cell the negative plate is made of cadmium. The electrolyte in each type of cell is a solution of potassium hydroxide which does not undergo any chemical change and thus the quantity can be reduced to a minimum. The plates are separated by insulating rods and assembled in steel containers which are then enclosed in a non-metallic crate to insulate the cells from one another. The average discharge p.d. of an alkaline cell is about 1.2 V.

Advantages of an alkaline cell (for example, a nickel-cadmium cell or a nickel-iron cell) over a lead-acid cell include:

> (i) More robust construction;
> (ii) Capable of withstanding heavy charging and discharging currents without damage;
> (iii) Has a longer life;
> (iv) For a given capacity is lighter in weight;
> (v) Can be left indefinitely in any state of charge or discharge without damage;
> (vi) is not self-discharging.

54

Disadvantages of an alkaline cell over a lead-acid cell include:

    (i) Is relatively more expensive;
    (ii) Requires more cells for a given emf;
    (iii) Has a higher internal resistance;
    (iv) Must be kept sealed;
    (v) Has a lower efficiency.

Alkaline cells may be used in extremes of temperature, in conditions where vibration is experienced or where duties require long idle periods or heavy discharge currents. Practical examples include traction and marine work, lighting in railway carriages, military portable radios and for starting diesel and petrol engines. However, the lead acid cell is the most common one in practical use.

11   The **capacity** of a cell is measured in ampere-hours (A h). A fully charged 50 A h battery rated for 10 h discharge can be discharged at a steady current of 5 A for 10 h, but if the load current is increased to 10 A, then the battery is discharged in 3–4 h, since the higher the discharge current, the lower is the effective capacity of the battery. Typical discharge characteristics for a lead-acid cell are shown in *Figure 9.6*.

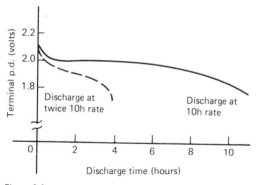

Figure 9.6

10 Capacitors and capacitance

1   **Electrostatics** is the branch of electricity which is concerned with the study of electrical charges at rest. An electrostatic field accompanies a static charge and this is utilised in the capacitor.

2   Charged bodies attract or repel each other depending on the nature of the charge. The rule is: **like charges repel, unlike charges attract.**

3   A **capacitor** is a device capable of storing electrical energy. *Figure 10.1* shows a capacitor consisting of a pair of parallel metal plates X and Y separated by an insulator, which could be air. Since the plates are electrical conductors each will contain a large number of mobile electrons. Since the plates are connected to a d.c. supply the electrons on plate X, which have a small negative charge, will be attracted to the positive pole of the supply and will be repelled from the negative pole of the supply on to plate Y. X will become positively charged due to its shortage of electrons whereas Y will have a negative charge due to its surplus of electrons.

   The difference in charge between the plates results in a p.d. existing between them, the flow of electrons dying away and ceasing when the p.d. between the plates equals the supply voltage. The plates are then said to be **charged** and there exists an **electric field** between them.

   Figure 10.2 shows a side view of the plates with the field represented by 'lines of electrical flux'. If the plates are discon-

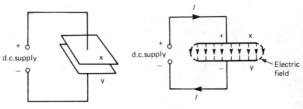

Figure 10.1   Figure 10.2

nected from the supply and connected together through a resistor the surplus of electrons on the negative plate will flow through the resistor to the positive plate. This is called **discharging**. The current flow decreases to zero as the charges on the plates reduce. The current flowing in the resistor causes it to liberate heat showing that **energy is stored in the electric field**.

4   From Section 7, para 2(viii), page 20, charge Q stored is given by:

$$Q = I \times t \text{ coulombs}$$

where I is the current in amperes and t the time in seconds.

5   A **dielectric** is an insulating medium separating charged surfaces.

6   Electric field strength, Electric force, or voltage gradient,

$$E = \frac{\text{p.d. across dielectric}}{\text{thickness of dielectric}}, \text{ i.e. } E = \frac{V}{d} \text{ volts/m.}$$

7   Electric flux density,

$$D = \frac{\text{charge}}{\text{Area of one plate}} \text{ i.e. } D = \frac{Q}{A} \text{ C/m}^2$$

8   Charge Q on a capacitor is proportional to the applied voltage V,

i.e. $Q \propto V$.

9    $Q = CV$ or $C = \frac{Q}{V}$, where the constant of proportionality, C, is the **capacitance**.

10   The unit of capacitance is the **farad F** (or more usually $\mu F = 10^{-6}$ F or pF $= 10^{-12}$ F), which is defined as the capacitance of a capacitor when a p.d. of one volt appears across the plates when charged with one coulomb.

11   Every system of electrical conductors possesses capacitance. For example, there is capacitance between the conductors of overhead transmission lines and also between the wires of a telephone cable. In these examples the capacitance is undesirable but has to be accepted, minimised or compensated for. There are other situations, such as in capacitors, where capacitance is a desirable property.

12   The ratio of electric flux density, D, to electric field strength, E, is called absolute permittivity, ε, of a dielectric.

Thus $\frac{D}{E} = \varepsilon$.

13 Permittivity of free space is a constant, given by

$\varepsilon_0 = 8.85 \times 10^{-12}$ F/m.

14 **Relative permittivity**,

$$\varepsilon_r = \frac{\text{flux density of the field in the dielectric}}{\text{flux density of the field in vacuum}}.$$

(ε_r has no units.) Examples of the values of ε_r include: air = 1, polythene = 2.3, mica = 3–7, glass = 5–10, ceramics = 6–1000.

15 Absolute permittivity, $\varepsilon = \varepsilon_0 \varepsilon_r$.

Thus $\dfrac{D}{E} = \varepsilon_0 \varepsilon_r$.

16 For a **parallel plate capacitor**, capacitance is proportional to area A, inversely proportional to the plate spacing (or dielectric thickness) d, and depends on the nature of the dielectric and the number of plates, n.

Capacitance, $C = \dfrac{\varepsilon_0 \varepsilon_r A (n - 1)}{d}$ F.

17 For n capacitors connected in parallel, the equivalent capacitance C_T is given by:

$C_T = C_1 + C_2 + C_3 + \ldots + C_n$ (similar to resistors connected in series)

Also total charge,

$Q_T = Q_1 + Q_2 + Q_3 + \ldots + Q_n$

18 For n capacitors connected in series, the equivalent capacitance C_T is given by:

$\dfrac{1}{C_T} = \dfrac{1}{C_1} + \dfrac{1}{C_2} + \dfrac{1}{C_3} + \ldots + \dfrac{1}{C_n}$ (similar to resistors connected in parallel)

The charge on each capacitor is the same when connected in series.

19 The maximum amount of field strength that a dielectric can withstand is called the **dielectric strength** of the material.

Dielectric strength, $E_{\text{MAX}} = \dfrac{V_{\text{MAX}}}{d}$ and $V_{\text{MAX}} = d \times E_{\text{MAX}}$.

20 The energy, W, stored by a capacitor is given by

$$W = \frac{1}{2} C V^2 \text{ joules.}$$

21 **Practical types of capacitor** are characterised by the material used for their dielectric. The main types include: variable air, mica, paper, ceramics, plastic and electrolytic.

(a) **Variable air capacitors**. These usually consist of two sets of metal plates (i.e. aluminium) one fixed, the other variable. The set of moving plates rotate on a spindle as shown by the end view in *Figure 10.3*. As the moving plates are rotated through half a revolution, the meshing, and therefore the capacitance, varies from a minimum to a maximum value. Variable air capacitors are used in radio and electronic circuits where very low losses are required, or where a variable capacitance is needed. The maximum value of such capacitors is between 500 pF and 1000 pF.

(b) **Mica capacitors** A typical older type construction is shown in *Figure 10.4*. Usually the whole capacitor is

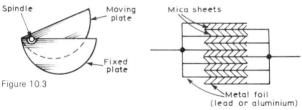

Figure 10.3

Figure 10.4

impregnated with wax and placed in a bakelite case. Mica is easily obtained in thin sheets and is a good insulator. However, mica is expensive and is not used in capacitors above about 0.1 μF. A modified form of mica capacitor is the silvered mica type. The mica is coated on both sides with a thin layer of silver which forms the plates. Capacitance is stable and less likely to change with age. Such capacitors have a constant capacitance with change of temperature, a high working voltage rating and a long service life and are used in high frequency circuits with fixed values of capacitance up to about 1000 pF.

(c) **Paper capacitors** A typical paper capacitor is shown in *Figure 10.5* where the length of the roll corresponds to the capacitance required. The whole is usually impregnated with oil or wax to exclude moisture, and then placed in a plastic or aluminium container for protection. Paper

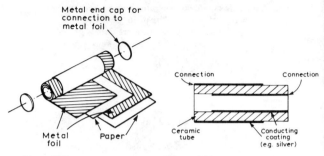

Figure 10.5

Figure 10.6

capacitors, up to about 1 μF, are made in various working
voltages. Disadvantages of paper capacitors include
variation in capacitance with temperature change and a
shorter service life than most other types of capacitor.

(d) **Ceramic capacitors** These are made in various forms,
each type of construction depending on the value of
capacitance required. For high values, a tube of ceramic
material is used as shown in the cross section of *Figure
10.6*. For smaller values the cup construction is used as
shown in *Figure 10.7*, and for still smaller values the disc
construction shown in *Figure 10.8* is used.

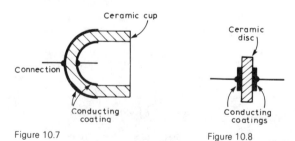

Figure 10.7

Figure 10.8

Certain ceramic materials have a very high permittivity and this enables capacitors of high capacitance
to be made which are of small physical size with a high
working voltage rating. Ceramic capacitors are available
in the range 1 pF to 0.1 μF and may be used in high

frequency electronic circuits subject to a wide range of temperature.

(e) **Plastic capacitors**   Some plastic materials such as polystyrene and Teflon can be used as dielectrics. Construction is similar to the paper capacitor but using a plastic film instead of paper. Plastic capacitors operate well under conditions of high temperature, provide a precise value of capacitance, a very long service life and high reliability.

(f) **Electrolytic capacitors**   Construction is similar to the paper capacitor with aluminium foil used for the plates and with a thick absorbent material, such as paper, impregnated with an electrolyte (ammonium borate), separating the plates. The finished capacitor is usually assembled in an aluminium container and hermetically sealed. Its operation depends on the formation of a thin aluminium oxide layer on the positive plate by electrolytic action when a suitable direct potential is maintained between the plates. This oxide layer is very thin and forms the dielectric. (The absorbent paper between the plates is a conductor and does not act as a dielectric.) Such capacitors **must only be used on d.c.** and must be connected with the correct polarity; if this is not done the capacitor will be destroyed since the oxide layer will be destroyed. Electrolytic capacitors are manufactured with working voltages from 6 V to 500 V, although accuracy is generally not very high. These capacitors possess a much larger capacitance than other types of capacitors of similar dimensions due to the oxide film being only a few microns thick. The fact that they can be used only on d.c. supplies limits their usefulness.

22   When a capacitor has been disconnected from the supply it may still be charged and it may retain this charge for some considerable time. Thus precautions must be taken to ensure that the capacitor is automatically discharged after the supply is switched off. This is done by connecting a high value resistor across the capacitor terminals.

11 Electromagnetism and magnetic circuits

Magnetism

1   A **permanent magnet** is a piece of ferromagnetic material (such as iron, nickel or cobalt) which has properties of attracting other pieces of these materials.

2   The area around a magnet is called the **magnetic field** and it is in this area that the effects of the **magnetic force** produced by the magnet can be detected.

3   The magnetic field of a bar magnet can be represented pictorially by the 'lines of force' (or lines of 'magnetic flux' as they are called) as shown in *Figure 11.1*. Such a field pattern can be

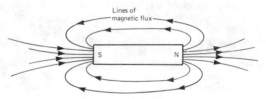

Figure 11.1

produced by placing iron filings in the vicinity of the magnet. The field direction at any point is taken as that in which the north-seeking pole of a compass needle points when suspended in the field. External to the magnet the direction of the field is north to south.

4   The laws of magnetic attraction and repulsion can be demonstrated by using two bar magnets. In *Figure 11.2(a)*, **with unlike poles adjacent, attraction occurs**. In *Figure 11.2(b)*, **with like poles adjacent, repulsion occurs**.

Electromagnetism

5   Magnetic fields are produced by electric currents as well as by permanent magnets. The field forms a circular pattern with the

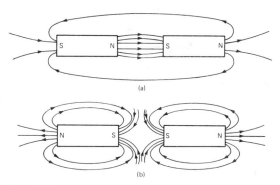

Figure 11.2

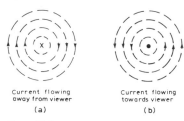

Current flowing
away from viewer
(a)

Current flowing
towards viewer
(b)

Figure 11.3

current carrying conductor at the centre. The effect is portrayed in *Figure 11.3* where the convention adopted is:

(a) current flowing **away** from the viewer is shown by $\oplus$ – can be thought of as the feathered end of the shaft of an arrow

(b) current flowing **towards** the viewer is shown by $\odot$ – can be thought of as the tip of an arrow.

6 The **direction** of the fields in *Figure 11.3* is remembered by the **screw rule** which states: *"If a normal right-hand thread screw is screwed along the conductor in the direction of the current, the direction of rotation of the screw is in the direction of the magnetic field'.*

7 A magnetic field produced by a long coil, or **solenoid**, is shown in *Figure 11.4* and is seen to be similar to that of a bar magnet shown in *Figure 11.1*. If the solenoid is wound on an iron bar an even stronger field is produced. The **direction** of the field produced by current *I* is determined by a compass and is remembered by either:

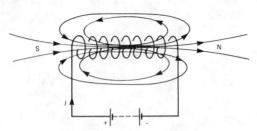

Figure 11.4

(a) the **screw rule**, which states that if a normal right hand thread screw is placed along the axis of the solenoid and is screwed in the direction of the current it moves in the direction of the magnetic field inside of the solenoid (i.e. points in the direction of the north pole), or

(b) the **grip rule**, which states that if the coil is gripped with the right hand with the fingers pointing in the direction of the current, then the thumb, outstretched parallel to the axis of the solenoid, points in the direction of the magnetic field inside the solenoid (i.e. points in the direction of the north pole).

8 An **electromagnet**, which is a solenoid wound on an iron core, provides the basis of many items of electrical equipment, examples including electric bells, relays and lifting magnets.

(i) A simple **electric bell circuit** is shown in *Figure 11.5*. When the switch S is closed a current passes through the coil. The iron-cored solenoid is energised, the soft iron armature is attracted to the electromagnet and the striker hits the gong. When the switch S is opened the coil becomes demagnetised and the spring steel strip pulls the armature back to its original position. This is the principle of operation of an electric bell.

(ii) A typical **relay circuit** connected to an alarm device is shown in *Figure 11.6*. When the switch S is closed a current passes through the coil and the iron-cored solenoid is energised. The hinged soft iron armature is attracted to the electromagnet and pushes against the two fixed contacts so that they are connected together, thus closing the electric circuit to be controlled – in this case, an alarm circuit. The alarm sounds for as long as the current flows in the coil.

(iii) A typical scrap-metal yard **lifting magnet** showing

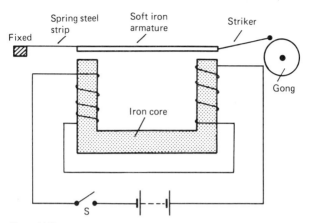

Figure 11.5

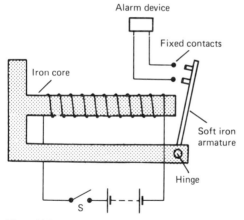

Figure 11.6

the plan and elevation is shown in *Figure 11.7*. When current is passed through the coil, the iron core becomes magnetised (i.e., an electromagnet) and this will attract to it other pieces of magnetic material. When the circuit is broken the iron core becomes demagnetised which releases the materials being lifted.

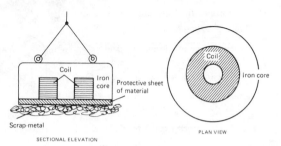

Scrap-metal

SECTIONAL ELEVATION

PLAN VIEW

Figure 11.7

9   (i) **Magnetic flux** is the amount of magnetic field (or the number of lines of force) produced by a magnetic source.
     (ii) The symbol for magnetic flux is Φ (Greek letter 'phi').
     (iii) The unit of magnetic flux is the **weber, Wb**.

10   (i) Magnetic flux density is the amount of flux passing through a defined area that is perpendicular to the direction of the flux.

$$\textbf{Magnetic flux density} = \frac{\textbf{magnetic flux}}{\textbf{area}}$$

     (ii) The symbol for magnetic flux density is B.
     (iii) The unit of magnetic flux density is the tesla, T, where $1\text{ T} = 1\text{ Wb/m}^2$.

     Hence $\boxed{B = \dfrac{\Phi}{A} \text{ tesla,}}$ where A is the area in m^2.

11   (i) If a current carrying conductor is placed in a magnetic field produced by permanent magnets then the fields due to the current carrying conductor and the permanent magnets intersect and cause a force to be exerted on the conductor. The force on the current carrying conductor in a magnetic field depends upon:

     (i) the intensity of the field, B teslas;
     (ii) the strength of the current, i amperes;
     (iii) the length of the conductor perpendicular to the magnetic field, l metres; and
     (iv) the directions of the field and the current.

     (ii) When the magnetic field, the current and the

66

conductor are mutually at right angles then:

Force $F = BIl$ newtons

(iii) When the conductor and the field are at an angle $\theta°$ to each other then:

Force $F = BIl \sin \theta$ newtons

(iv) Since when the magnetic field, current and conductor are mutually at right angles, $F = BIl$, the magnetic flux density B may be defined by $B = \dfrac{F}{Il}$, i.e. a field intensity of 1 T is exerted on 1 m of a conductor when the conductor carries a current of 1 A.

12 If the current-carrying conductor shown in *Figure 11.3(a)* is placed in the magnetic field shown in *Figure 11.8(a)*, then the two fields interact and cause a force to be exerted on the conductor as shown in *Figure 11.8(b)*. The field is strengthened above the

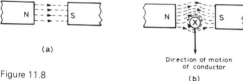

(a)

Direction of motion
of conductor

Figure 11.8

(b)

conductor and weakened below, thus tending to move the conductor downwards. This is the basic principle of operation of the electric motor (see para 14) and the moving coil instrument (see page 184).

13 The direction of the force exerted on a conductor can be predetermined by using **Fleming's left-hand rule** (often called the motor rule), which states:

'*Let the thumb, first finger and second finger of the left-hand be extended such that they are all at right angles to each other, as shown in Figure 11.9. If the first finger points in the direction of the magnetic field, the second finger points in the direction of the current, then the thumb will point in the direction of the motion of the conductor.*'

Summarising:

First finger   – Field
SeCond finger   – Current
ThuMb   – Motion

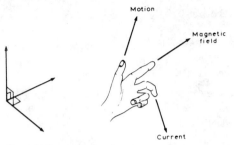

Figure 11.9

Principle of operation of a d.c. motor

14   A rectangular coil which is free to rotate about a fixed axis is shown placed inside a magnetic field produced by permanent magnets in *Figure 11.10*. A direct current is fed into the coil via

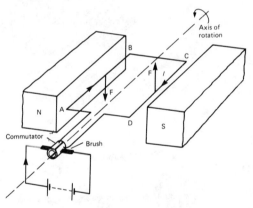

Figure 11.10

carbon brushes bearing on a commutator, which consists of a metal ring split into two halves separated by insulation. When current flows in the coil, a magnetic field is set up around the coil which interacts with the magnetic field produced by the magnets. This causes a force F to be exerted on the current carrying conductor, which, by Fleming's left-hand rule (see para 13) is downwards between points A and B and upwards between C and D for the current direction shown. This causes a torque and the coil rotates anticlockwise.

When the coil has turned through 90° from the position shown in *Figure 11.10* the brushes connected to the positive and negative terminals of the supply make contact with different halves of the commutator ring, thus reversing the direction of the current flow in the conductor. If the current is not reversed and the coil rotates past this position the forces acting on it change direction and it rotates in the opposite direction thus never making more than half a revolution. The current direction is reversed every time the coil swings through the vertical position and thus the coil rotates anticlockwise for as long as the current flows. This is the principle of operation of a d.c. motor which is thus a device that takes in electrical energy and converts it into mechanical energy.

Magnetic circuits

15 **Magnetomotive force (mmf)**, $F_m = NI$ **ampere-turns (At)**, where N = number of conductors (or turns) and I = current in amperes.

Since 'turns' has no units, the SI unit of mmf is the ampere, but to avoid any possible confusion 'ampere-turns', (A t) are used in this chapter.

16 **Magnetic field strength**, or **magnetising force**

$$H = \frac{NI}{l} \text{ A t/m}$$

where l = mean length of flux path in metres.

Hence, mmf = NI = Hl At.

17 **For air, or any non-magnetic medium**, the ratio of magnetic flux density to magnetising force is a constant, i.e.

$$\frac{B}{H} = \text{a constant.}$$

This constant is μ_0, the permeability of free space (or the magnetic space constant) and is equal to $4\pi \times 10^{-7}$ H/m.

Hence, for a non-magnetic medium, $\frac{B}{H} = \mu_0$

18 For ferromagnetic mediums:

$$\frac{B}{H} = \mu_0 \mu_r,$$

where μ_r is the relative permeability, and is defined as

$$\frac{\text{flux density in material}}{\text{flux density in air}}$$

Its value varies with the type of magnetic material and since μ_r is a ratio of flux densities, it has no units. From its definition, μ_r for air is 1.

19    $\mu_0\mu_r = \mu$, called the **absolute permeability**.

20   By plotting measured values of flux density B against magnetic field strength H, a **magnetisation curve** (or B–H curve) is produced. For non-magnetic materials this is a straight line. Typical curves for four magnetic materials are shown in *Figure 11.11*.

21   The relative permeability of a ferromagnetic material is

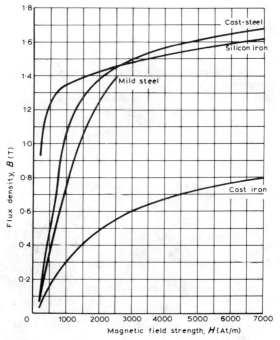

Figure 11.11

proportional to the slope of the $B-H$ curve and thus varies with the magnetic field strength. The approximate range of values of relative permeability μ_r for some common magnetic materials are:

Cast iron $\mu_r = 100 - 250$; Mild steel $\mu_r = 200 - 800$
Silicon iron $\mu_r = 1000 - 5000$; Cast steel $\mu_r = 300 - 900$
Mumetal $\mu_r = 200 - 5000$; Stalloy $\mu_r = 500 - 6000$

22 The 'magnetic resistance' of a magnetic circuit to the presence of magnetic flux is called **reluctance**. The symbol for reluctance is S (or R_m).

23 Reluctance

$$S = \frac{\text{mmf}}{\Phi} = \frac{NI}{\Phi} = \frac{Hl}{BA} = \frac{l}{\frac{B}{H}A} = \frac{l}{\mu_0 \mu_r A}$$

24 The unit of reluctance is $1/H$ (or H^{-1}) or At/Wb.

25 For a series magnetic circuit having n parts, the total reluctance S is given by: $S = S_1 + S_2 + \ldots + S_n$. (This is similar to resistors connected in series in an electrical circuit.)

Comparison between electrical and magnetic quantities:

26

Electric circuit	Magnetic circuit
emf E (V)	mmf F_m (At)
current I (A)	flux Φ (Wb)
resistance R (Ω)	reluctance S (H^{-1})
$I = \dfrac{E}{R}$	$\Phi = \dfrac{\text{mmf}}{S}$
$R = \dfrac{\rho l}{A}$	$S = \dfrac{l}{\mu_0 \mu_r A}$

27 Ferromagnetic materials have a low reluctance and can be used as **magnetic screens** to prevent magnetic fields affecting materials within the screen.

28 **Hysteresis** is the 'lagging' effect of flux density B whenever there are changes in the magnetic field strength H.

29   When an initially unmagnetised ferromagnetic material is subjected to a varying magnetic field strength H, the flux density B produced in the material varies as shown in *Figure 11.12*, the arrows indicating the direction of the cycle. *Figure 11.12* is known as the hysteresis loop.

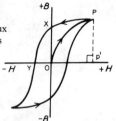

Figure 11.12

OX = **residual flux density** or **remanence**
OY = **coercive force**
PP = **saturation flux density**

30   Hysteresis results in a dissipation of energy which appears as a heating of the magnetic material. The **energy loss** associated with hysteresis is proportional to the areas of the hysteresis loop.

31   The **area** of a hysteresis loop varies with the type of material. The area, and thus the energy loss, is much greater for hard materials than for soft materials.

12 Electromagnetic induction and inductance

1   When a conductor is moved across a magnetic field, an electromotive force (emf) is produced in the conductor. If the conductor forms part of a closed circuit then the emf produced causes an electric current to flow round the circuit. Hence an emf (and thus current), is 'induced' in the conductor as a result of its movement across the magnetic field. This effect is known as **'electromagnetic induction'**.

2   **Faraday's laws of electromagnetic induction** state:

> (i) *'An induced emf is set up whenever the magnetic field linking that circuit changes.'*
> (ii) *'The magnitude of the induced emf in any circuit is proportional to the rate of change of the magnetic flux linking the circuit.'*

3   **Lenz's law** states:

> *'The direction of an induced emf is always such as to oppose the effect producing it.'*

4   An alternative method to Lenz's law of determining relative directions is given by **Fleming's <u>R</u>ight-hand rule** (often called the gene<u>R</u>ator rule) which states:

> *'Let the thumb, first finger and second finger on the right hand be extended such that they are all at right angles to each other, as shown in Figure 12.1. If the first finger points in the direction of the magnetic field the thumb points in the direction of motion of the conductor relative to the magnetic field, then the second finger will point in the direction of the induced emf.'*

Summarising:

<u>F</u>irst finger	–	<u>F</u>ield
Thu<u>M</u>b	–	<u>M</u>otion
S<u>E</u>cond finger	–	<u>E</u>mf

5   In a **generator**, conductors forming an electric circuit are made to move through a magnetic field. By Faraday's law an emf is induced in the conductors and thus a source of emf is created. A

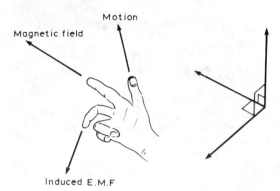

Figure 12.1

generator converts mechanical energy into electrical energy. (The action of a simple a.c. generator is described in para 2, page 78.)

6   The induced emf E set up between the ends of the conductor shown in *Figure 12.2* is given by: $E = Blv$ **volts**, where B, the flux

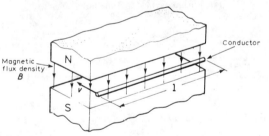

Figure 12.2

density is measured in teslas, l, the length of conductor in the magnetic field is measured in metres, and v, the conductor velocity, is measured in metres per second. If the conductor moves at an angle $\theta°$ to the magnetic field (instead of at $90°$ as assumed above) then $E = Blv \sin \theta$.

7   **Inductance** is the name given to the property of a circuit whereby there is an emf induced into the circuit by the change of flux linkages produced by a current change.

(i) When the emf is induced in the same circuit as that in which the current is changing, the property is called **self inductance**, L;

(ii) When the emf is induced in a circuit by a change of flux due to current changing in an adjacent circuit, the property is called **mutual inductance**, M.

8   The unit of inductance is the **henry, H**.

'A circuit has an inductance of one henry when an emf of one volt is induced in it by a current changing at the rate of one ampere per second.'

9   A component called an **inductor** is used when the property of inductance is required in a circuit. The basic form of an inductor is simply a coil of wire. Factors which affect the inductance of an inductor include:

(i) the number of turns of wire – the more turns the higher the inductance;

(ii) the cross-sectional area of the coil of wire – the greater the cross-sectional area the higher the inductance;

(iii) the presence of a magnetic core – when the coil is wound on an iron core the same current sets up a more concentrated magnetic field and the inductance is increased;

(iv) the way the turns are arranged – a short thick coil of wire has a higher inductance than a long thin one.

10   Two examples of practical inductors are shown in *Figure 12.3* and the standard electrical circuit diagram symbol for air-cored and iron-cored inductors are shown in *Figure 12.4*.

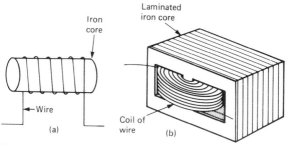

Figure 12.3

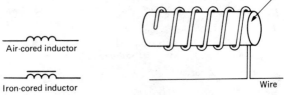

Air-cored inductor

Iron-cored inductor

Figure 12.4

Insulator

Wire

Figure 12.5

An iron-cored inductor is often called a **choke** since, when used in a.c. circuits it has a choking effect, limiting the current flowing through it.

11 Inductance is often undesirable in a circuit. To reduce inductance to a minimum the wire may be bent back on itself as shown in *Figure 12.5* so that the magnetising effect of one conductor is neutralised by that of the adjacent conductor. The wire may be coiled around an insulator as shown without increasing the inductance. Standard resistors may be non-inductively wound in this manner.

12 An inductor possesses an ability to store energy. The **energy stored**, W, in the magnetic field of an inductor is given by:

$$W = \frac{1}{2}LI^2 \text{ joules.}$$

13 (i) Induced emf in a coil of N turns, $E = N\left(\dfrac{\Delta\Phi}{t}\right)$ volts,

where $\Delta\Phi$ is the change in flux, in Webers, and t is the time taken for the flux to change, in seconds.

(ii) Induced emf in a coil of inductance L henrys

$E = L\left(\dfrac{\Delta I}{t}\right)$ volts, where ΔI is the change in current, in amperes, and t is the time taken for the current to change, in seconds.

14 If a current changing from 0 to I amperes, produces a flux change from 0 to Φ webers, then $\Delta I = I$ and $\Delta\Phi = \Phi$. Then, from para 13, induced emf $E = \dfrac{N\Phi}{t} = \dfrac{LI}{t}$, from which

inductance of coil, $L = \dfrac{N\Phi}{I}$ **henrys**.

15   From para 23, page 71,

reluctance $S = \dfrac{\text{mmf}}{\text{flux}} = \dfrac{NI}{\Phi}$, from which $\Phi = \dfrac{NI}{S}$.

Hence, inductance of coil,

$$L = \frac{N\Phi}{I} = \frac{N}{I}\left(\frac{NI}{S}\right) = \frac{N^2}{S}, \text{ i.e. } L \alpha N^2.$$

16   Mutually induced emf in the second coil, $E_2 = N\left(\dfrac{\Delta I_1}{t}\right)$ volts,

where M is the mutual inductance between two coils, in henrys, ΔI_1 is the change in current in the first coil, in amperes, and t is the time the current takes to change in the first coil, in seconds. A transformer is a device which uses the phenomenon of mutual inductance to change the value of alternating voltages. (See chapter 21, page 156.)

13 Alternating currents and voltages

1   Electricity is produced by generators at power stations and then distributed by a vast network of transmission lines (called the National Grid system) to industry and for domestic use. It is easier and cheaper to generate alternating current (a.c.) than direct current (d.c.) and a.c. is more conveniently distributed than d.c. since its voltage can be readily altered using transformers. Whenever d.c. is needed in preference to a.c. devices called rectifiers are used for conversion (see paragraphs 19 to 23).

2   Let a single turn coil be free to rotate at constant angular velocity ω symmetrically between the poles of a magnet system as shown in *Figure 13.1*. An emf is generated in the coil (from

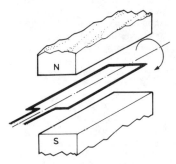

Figure 13.1

Faraday's law) which varies in magnitude and reverses its direction at regular intervals. The reason for this is shown in *Figure 13.2*. In positions (a), (e) and (i) the conductors of the loop are effectively moving along the magnetic field, no flux is cut and hence no emf is induced. In position (c) maximum flux is cut and hence maximum emf is induced. In position (g), maximum flux is cut and hence maximum emf is again induced. However, using Fleming's right-hand rule, the induced emf is in the opposite direction to that in

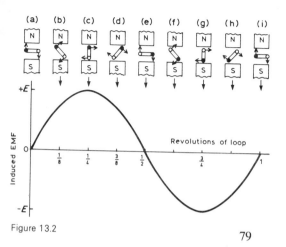

Figure 13.2

position (c) and is thus shown as $-E$. In positions (b), (d), (f) and (h) some flux is cut and hence some emf is induced. If all such positions of the coil are considered, in one revolution of the coil, one cycle of alternating emf is produced as shown. This is the principle of operation of the **a.c. generator** (i.e. the **alternator**).

3   If values of quantities which vary with time t are plotted to a base of time, the resulting graph is called a **waveform**. Some typical waveforms are shown in *Figure 13.3*. Waveforms (a) and (b) are **unidirectional waveforms,** for, although they vary considerably with time, they flow in one direction only (i.e. they do not cross the time axis and become negative). Waveforms (c) to (g) are called **alternating waveforms** since their quantities are continually changing in direction (i.e. alternately positive and negative).

4   A waveform of the type shown in *Figure 13.3(g)* is called a **sine wave**. It is the shape of the waveform of emf produced by an alternator and thus the mains electricity supply is of 'sinusoidal' form.

5   One complete series of values is called a **cycle** (i.e. from O to P in *Figure 13.3(g)*).

6   The time taken for an alternating quantity to complete one cycle is called the **period** or the **periodic time,** T, of the waveform.

7   The number of cycles completed in one second is called the **frequency,** f, of the supply and is measured in hertz, Hz. The

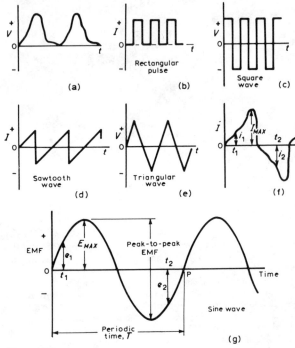

Figure 13.3

standard frequency of the electricity supply in Great Britain is 50 Hz.

$$T = \frac{1}{f} \text{ or } f = \frac{1}{T}.$$

8   **Instantaneous values** are the values of the alternating quantities at any instant of time. They are represented by small letters, i, v, e etc. (see *Figure 13.3(f)* and *(g)*).

9   The largest value reached in a half cycle is called the **peak value** or the **maximum value** or the **crest value** or the **amplitude** of the waveform. Such values are represented by V_{MAX}, I_{MAX} etc. (see *Figure 13.3(f)* and *(g)*). A **peak-to-peak value** of emf is shown in *Figure 13.3(g)* and is the difference between the maximum and minimum values in a cycle.

10   The **average or mean value** of a symmetrical alternating quantity, (such as a sinewave), is the average value measured over a half cycle, (since over a complete cycle the average value is zero).

$$\text{Average or mean value} = \frac{\text{area under the curve}}{\text{length of base}}$$

The area under the curve is found by approximate methods such as the trapezoidal rule, the mid-ordinate rule or Simpson's rule. Average values are represented by V_{AV}, I_{AV}, etc.

For a sine wave, average value $= 0.637 \times$ maximum value $\left(\text{i.e. } \dfrac{2}{\pi} \times \text{maximum value} \right)$.

11   The **effective value** of an alternating current is that current which will produce the same heating effect as an equivalent direct current. The effective value is called the **root mean square (r.m.s.) value** and whenever an alternating quantity is given, it is assumed to be the r.m.s. value. For example, the domestic mains supply in Great Britain is 240 V and is assumed to mean '240 V r.m.s.'. The symbols used for r.m.s. values are I, V, E, etc. For a non-sinusoidal waveform as shown in *Figure 13.4*, the r.m.s. value is given by:

$$I = \sqrt{\left\{ \frac{i_1^2 + i_2^2 + \dots + i_n^2}{n} \right\}}$$

where n is the number of intervals used.

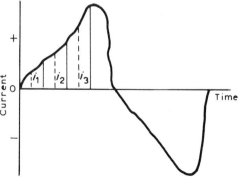

Figure 13.4

81

For a sine wave, r.m.s. value = 0.707 × maximum value

$$\left(\text{i.e. } \frac{1}{\sqrt{2}} \times \text{maximum value}\right).$$

12 (a) **Form factor** = $\dfrac{\text{r.m.s. value}}{\text{average value}}$.

  For a sine wave, form factor = 1.11

  (b) **Peak factor** = $\dfrac{\text{maximum value}}{\text{r.m.s. value}}$.

  For a sine wave, peak factor = 1.41.

The values of form and peak factors give an indication of the shape of waveforms.

13 In *Figure 13.5*, OA represents a vector that is free to rotate anticlockwise about 0 at an angular velocity of ω rad/s. A rotating

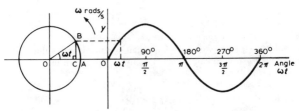

Figure 13.5

vector is known as a *phasor*. After time t seconds the vector OA has turned through an angle ωt. If the line BC is constructed perpendicular to OA as shown, then

$$\sin \omega t = \frac{BC}{OB}, \text{ i.e. } BC = OB \sin \omega t.$$

If all such vertical components are projected on to a graph of y against angle ωt (in radians), a sine curve results of maximum value OA. Any quantity which varies sinusoidally can thus be represented as a phasor.

14 A sine curve may not always start at 0°. To show this a periodic function is represented by $y = \sin(\omega t \pm \phi)$, where ϕ is a phase (or angle) difference compared with $y = \sin \omega t$. In *Figure 13.6(a)*, $y_2 = \sin(\omega t + \phi)$ starts ϕ radians earlier than $y_1 = \sin \omega t$ and is thus said to lead y_1 by ϕ radians. Phasors y_1 and y_2 are shown in

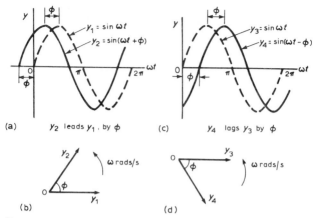

Figure 13.6

Figure 13.6(b) at the time when $t=0$. In *Figure 13.6(c)*, $y_4=\sin(\omega t-\phi)$ starts ϕ radians later than $y_3=\sin\omega t$ and is thus said to lag y_3 by ϕ radians. Phasors y_3 and y_4 are shown in *Figure 13.6(d)* at the time when $t=0$.

15 Given the general sinusoidal voltage, $V=V_{\text{MAX}}\sin(\omega t\pm\phi)$, then

(i) Amplitude or maximum value$=V_{\text{MAX}}$;

(ii) Peak to peak value$=2\ V_{\text{MAX}}$;

(iii) Angular velocity$=\omega$ rad/s.

(iv) Periodic time, $T=\dfrac{2\pi}{\omega}$ seconds.

(v) Frequency, $f=\dfrac{\omega}{2\pi}$ Hz (hence $\omega=2\pi f$).

(vi) $\phi=$ angle of lag or lead (compared with $v=V_{\text{MAX}}\sin\omega t$).

16 The **resultant of the addition (or subtraction) of two sinusoidal quantities** may be determined either:

(a) by plotting the periodic function graphically, or

(b) by resolution of phasors by drawing or calculation.

For example, currents $i_1=20\sin\omega t$ and $i_2=10\sin\left(\omega t+\dfrac{\pi}{3}\right)$ are shown plotted in *Figure 13.7*.

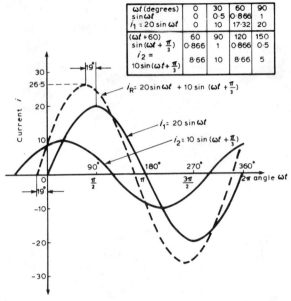

ωt (degrees)	0	30	60	90
$\sin \omega t$	0	0.5	0.866	1
$i_1 = 20 \sin \omega t$	0	10	17.32	20
$(\omega t + 60)$	60	90	120	150
$\sin(\omega t + \frac{\pi}{3})$	0.866	1	0.866	0.5
$i_2 = 10\sin(\omega t + \frac{\pi}{3})$	8.66	10	8.66	5

Figure 13.7

To determine the resultant of $i_1 + i_2$, ordinates of i_1 and i_2 are added at, say, $15°$ intervals. For example:

at $30°$, $i_1 + i_2 = 10 + 10 = 20$ A
at $60°$, $i_1 + i_2 = 8.7 + 17.3 = 26$ A
at $150°$, $i_1 + i_2 = 10 + (-5) = 5$ A, and so on.

The resultant waveform for $i_1 + i_2$ is shown by the broken line in *Figure 13.7*. It has the same period, and hence frequency, as i_1 and i_2. The amplitude or peak value is 26.5 A. The resultant waveform leads $i_1 = 20 \sin \omega t$ by $19°$, i.e., $\left(19 \times \dfrac{\pi}{180}\right)$ rad $= 0.332$ rad. Hence the sinusoidal expression for the resultant $i_1 + i_2$ is given by:

$$i_R = i_1 + i_2 = 26.5 \sin(\omega t + 0.332) \text{ A}.$$

The relative positions of i_1 and i_2 at time $= 0$ are shown as phasors in *Figure 13.8(a)*. The phasor diagram in *Figure 13.8(b)* shows the resultant i_R, and i_R is measured as 26 A and angle ϕ as $19°$ (i.e.

84

0.33 rad) leading i_1. Hence, by drawing, $i_R = 26 \sin(\omega t + 0.33)$ A.
From *Figure 13.8(b)*, by the cosine rule:

$$i_R^2 = 20^2 + 10^2 - 2(20)(10) \cos 120°,$$

from which, $i_R = 26.46$ A. By the sine rule:

$$\frac{10}{\sin \phi} = \frac{26.46}{\sin 120°}$$

from which,

$\phi = 19° 10'$ (i.e., 0.333 rad).

Hence, by calculation,

$i_R = 26.46 \sin(\omega t + 0.333)$ A.

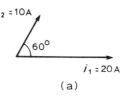

(a)

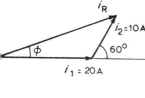

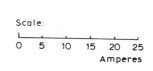

(b)

Figure 13.8

17   When a sinusoidal voltage is applied to a purely resistive circuit of resistance R, the voltage and current waveforms are in phase and $I = \dfrac{V}{R}$ (exactly as in d.c. circuit). V and I are r.m.s. values.

18   For an a.c. resistive circuit, power $P = VI = I^2 R = \dfrac{V^2}{R}$ watts (exactly as in a d.c. circuit). V and I are r.m.s. values.

19   The process of obtaining unidirectional currents and voltages from alternating currents and voltages is called **rectification**. Automatic switching in circuits is carried out by devices called diodes (see page 115).

20   Using a single diode, as shown in *Figure 13.9*, **half-wave rectification** is obtained. When P is sufficiently positive with respect to Q, diode D is switched on and current i flows.

When P is negative with respect to Q, diode D is switched off. Transformer T isolates the equipment from direct connection with the mains supply and enables the mains voltage to be changed.

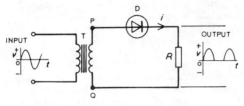

Figure 13.9

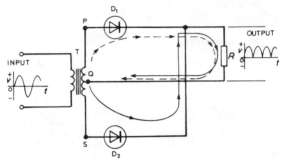

Figure 13.10

— — → — Current flow when P is positive W.R.T. Q

———→ — Current flow when Q is positive W.R.T. P

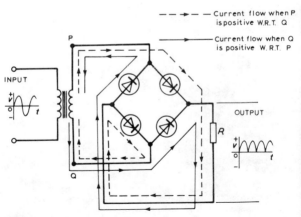

Figure 13.11

21　Two diodes may be used as shown in *Figure 13.10* to obtain **full wave rectification**. A centre-tapped transformer T is used. When P is sufficiently positive with respective to Q, diode D_1 conducts and current flows (shown by the broken line in *Figure 13.8*). When S is positive with respect to Q, diode D_2 conducts and current flows (shown by continuous line in *Figure 13.10*). The current flowing in R is in the same direction for both half cycles of the input. The output waveform is thus as shown in *Figure 13.10*.

22　Four diodes may be used in a **bridge rectifier circuit**, as shown in *Figure 13.11* to obtain **full wave rectification**. As for the rectifier shown in *Figure 13.10*, the current flowing in R is in the same direction for both half cycles of the input giving the output waveform shown.

23　To **smooth the output** of the rectifiers described above, capacitors having a large capacitance may be connected across the load resistors R. The effect of this is shown on the output in *Figure 13.12*.

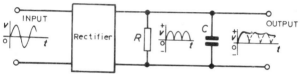

Figure 13.12

14 Single phase series a.c. circuits

1   In a **purely resistive a.c. circuit**, the current I_R and applied voltage V_R are in phase, see *Figure 14.1(a)*.

2   In a **purely inductive a.c. circuit**, the current I_L **lags** the applied voltage V_L by 90° (i.e. $\pi/2$ rads), see *Figure 14.1(b)*.

3   In a **purely capacitive a.c. circuit**, the current I_C **leads** the applied voltage V_C by 90° (i.e. $\pi/2$ rads), see *Figure 14.1(c)*.

4   In a purely inductive circuit the opposition to the flow of alternating current is called the **inductive reactance**, X_L.

$$X_L = \frac{V_L}{I_L} = 2\pi fL \ \Omega,$$

where f is the supply frequency in hertz and L is the inductance in henry's.

X_L is proportional to f as shown in *Figure 14.2(a)*.

5   In a purely capacitive circuit the opposition to the flow of alternating current is called the **capacitive reactance**, X_C.

$$X_C = \frac{V_C}{I_C} = \frac{1}{2\pi fC} \ \Omega,$$

where C is the capacitance in farads. X_C varies with f as shown in *Figure 14.2(b)*.

6   In an a.c. series circuit containing inductance L and resistance R, the applied voltage V is the phasor sum of V_R and V_L (see *Figure 14.3(a)*) and thus the current I lags the applied voltage V by an angle lying between 0° and 90° (depending on the values of V_R and V_L), shown as angle ϕ. In any a.c. series circuit the current is common to each component and is thus taken as the reference phasor.

7   In an a.c. series circuit containing capacitance C and resistance R, the applied voltage V is the phasor sum of V_R and V_C (see *Figure 14.3(b)*) and thus the current I leads the applied voltage V by an angle lying between 0° and 90° (depending on the values of V_R and V_C), shown as angle α.

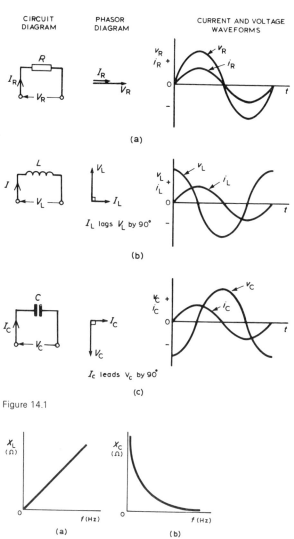

CIRCUIT DIAGRAM | PHASOR DIAGRAM | CURRENT AND VOLTAGE WAVEFORMS

(a)

(b)

I_L lags V_L by 90°

(c)

I_C leads V_C by 90°

Figure 14.1

(a)

(b)

Figure 14.2

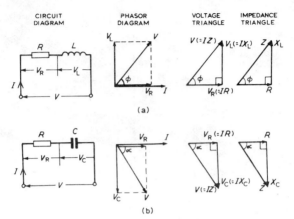

CIRCUIT DIAGRAM PHASOR DIAGRAM VOLTAGE TRIANGLE IMPEDANCE TRIANGLE

(a)

(b)

Figure 14.3

8 In an a.c. circuit, the ratio

$$\frac{\text{applied voltage } V}{\text{current } I}$$

is called the **impedance** Z,

i.e. $Z = \dfrac{V}{I}\ \mathbf{\Omega}.$

9 From the phasor diagrams of *Figure 14.3*, the '**voltage triangles**' are derived.

(a) For the *R-L* circuits:

$V = \sqrt{(V_R^2 + V_L^2)}$ (by Pythagoras' theorem), and

$\tan \phi = \dfrac{V_L}{V_R}$ (by trigonometric ratios)

(b) For the *R-C* circuit:

$V = \sqrt{(V_R^2 + V_C^2)}$, and

$\tan \alpha = \dfrac{V_C}{V_R}$

10 If each side of the voltage triangles in *Figure 14.3* is divided by current I then the '**impedance triangles**' are derived.

(a) For the R-L circuit: $Z = \sqrt{(R^2 + X_L^2)}$

$\tan \phi = \dfrac{X_L}{R}$, $\sin \phi = \dfrac{X_L}{Z}$ and $\cos \phi = \dfrac{R}{Z}$

(b) For the R-C circuit: $Z = \sqrt{(R^2 + X_C^2)}$

$\tan \alpha = \dfrac{X_C}{R}$, $\sin \alpha = \dfrac{X_C}{Z}$ and $\cos \alpha = \dfrac{R}{Z}$

11 In an a.c. series circuit containing resistance R, inductance L and capacitance C, the applied voltage V is the phasor sum of V_R, V_L and V_C (see *Figure 14.4*). V_L and V_C are anti-phase and there are three phasor diagrams possible — each depending on the relative values of V_L and V_C.

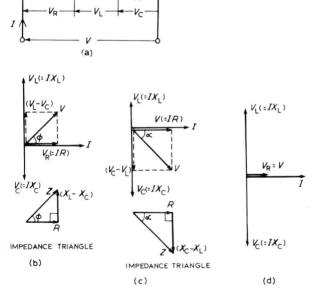

Figure 14.4

12 When $X_L > X_C$ (see *Figure 14.4(b)*):

$$Z = \sqrt{[R^2 + (X_L - X_C)^2]}$$

and $\tan \phi = \dfrac{(X_L - X_C)}{R}$

13 When $X_C > X_L$ (see *Figure 14.4(c)*):

$$Z = \sqrt{[R^2 + (X_C - X_L)^2]}$$

and $\tan \alpha = \dfrac{(X_C - X_L)}{R}$

14 When $X_L = X_C$ (see *Figure 14.4(d)*), the applied voltage V and the current I are in phase. This effect is called **series resonance**. At resonance:

(i) $V_L = V_C$

(ii) $Z = R$ (i.e. the minimum circuit impedance possible in an *L-C-R* circuit).

(iii) $I = \dfrac{V}{R}$ (i.e. the maximum current possible in an *L-C-R* circuit).

(iv) Since $X_L = X_C$, then $2\pi f_r L = \dfrac{1}{2\pi f_r C}$,

from which, $f_r = \dfrac{1}{2\pi \sqrt{(LC)}}$ Hz, where f_r is the resonant frequency.

15 At resonance, if R is small compared with X_L and X_C, it is possible for V_L and V_C to have voltages many times greater than the supply voltage V (see *Figure 14.4(d)*).

Voltage magnification at resonance

$$= \frac{\text{voltage across } L \text{ (or } C)}{\text{supply voltage } V}$$

This ratio is a measure of the quality of a circuit (as a resonator or tuning device) and is called the **Q-factor**.

Hence **Q-factor** $= \dfrac{V_L}{V} = \dfrac{I X_L}{IR} = \dfrac{X_L}{R} = \dfrac{2\pi f_r L}{R}$

$$\left(\text{Alternatively, Q-factor} = \frac{V_C}{V} = \frac{I X_C}{IR} = \frac{X_C}{R} = \frac{1}{2\pi f_r CR} \right)$$

At resonance $f_r = \dfrac{1}{2\pi \sqrt{(LC)}}$ i.e. $2\pi f_r = \dfrac{1}{\sqrt{(LC)}}$

Hence **Q-factor** $= \dfrac{2\pi f_r L}{R} = \dfrac{1}{\sqrt{(LC)}}\left(\dfrac{L}{R}\right) = \dfrac{1}{R}\sqrt{\left(\dfrac{L}{C}\right)}$

16   (a)   The advantages of a high Q-factor in a series high
        frequency circuit is that such circuits need to be
        'selective' to signals at the resonant frequency and must
        be less selective to other frequencies.

    (b)   The disadvantage of a high Q-factor in a series power
        circuit is that it can lead to dangerously high voltages
        across the insulation and may lead to electrical
        breakdown.

17   For series-connected impedances the total circuit impedance
can be represented as a single *L-C-R* circuit by combining all
values of resistance together, all values of inductance together and
all values of capacitance together (remembering that for series
connected capacitors $1/C = 1/C_1 + 1/C_2 + \cdots$). For example, the
circuit of *Figure 14.5(a)* showing three impedances has an
equivalent circuit of *Figure 14.5(b)*.

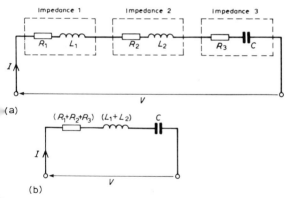

(a)

(b)

Figure 14.5

18   (a)   For a purely resistive a.c. circuit, the average power
        dissipated, *P*, is given by: $P = VI = I^2R = V^2/R$ watts
        (*V* and *I* being r.m.s. values). See *Figure 14.6(a)*.

    (b)   For a purely inductive a.c. circuit, the average power is
        zero. See *Figure 14.6(b)*.

    (c)   For a purely capacitive a.c. circuit, the average power is
        zero. See *Figure 14.6(c)*.

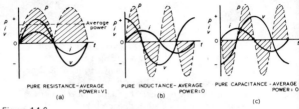

Figure 14.6

In *Figure 14.6(a)–(c)*, the value of power at any instant is given by the product of the voltage and current at that instant, i.e. the instantaneous power, $p = vi$, as shown by the broken lines.

19   *Figure 14.7* shows current and voltage waveforms for an *R-L* circuit where the current lags the voltage by angle ϕ. The

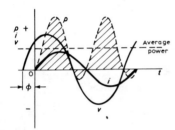

Figure 14.7

waveform for power (where $p = vi$) is shown by the broken line, and its shape, and hence average power, depends on the value of angle ϕ.

For an *R-L*, *R-C* or *L-C-R* series a.c. circuit, the average power *P* is given by:

$P = VI \cos \phi$ **watts** or $P = I^2 R$ **watts** (*V* and *I* being r.m.s. values).

20   *Figure 14.8(a)* shows a phasor diagram in which the current *I* lags the applied voltage *V* by angle ϕ. The horizontal component of *V* is $V \cos \phi$ and the vertical component of *V* is $V \sin \phi$. If each of the voltage phasors is multiplied by *I*, *Figure 14.8(b)* is obtained and is known as the '**power triangle**'.

Apparent power, $S = VI$ voltamperes (VA)
True or active power, $P = VI \cos \phi$ watts (W)
Reactive power, $Q = VI \sin \phi$ reactive voltamperes (VAr)

94

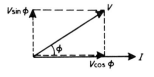

(a) PHASOR DIAGRAM

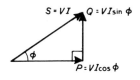

(b) POWER TRIANGLE

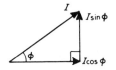

(c) CURRENT TRIANGLE

Figure 14.8

21   If each of the phasors of the power triangle of *Figure 14.8(b)* is divided by V, *Figure 14.8(c)* is obtained and is known as the '**current triangle**'. The horizontal component of current, $I \cos \phi$, is called the **active** or the **in-phase component**. The vertical component of current, $I \sin \phi$, is called the **reactive** or the **quadrature component**.

22   Power factor $= \dfrac{\text{True power } P}{\text{Apparent power } S}$

For sinusoidal voltages and currents,

$$\text{power factor} = \frac{P}{S} = \frac{VI \cos \phi}{VI}$$

i.e. **p.f.** $= \cos \phi = \dfrac{R}{Z}$ (from *Figure 14.3*).

(The relationships stated in paras 20 to 22 are also true when current I leads voltage V.)

15 Single phase parallel a.c. circuits

1   In parallel circuits, such as those shown in *Figure 15.1*, the voltage is common to each branch of the network and is thus taken as the reference phasor when drawing phasor diagrams.

2   **R-L parallel circuit**

   In the two branch parallel circuit containing resistance R and inductance L shown in *Figure 15.1(a)*, the current flowing in the resistance, I_R, is in-phase with the supply voltage V and the current flowing in the inductance, I_L, lags the supply voltage by 90°. The supply current I is the phasor sum of I_R and I_L and thus the current I lags the applied voltage V by an angle lying between 0° and 90° (depending on the values of I_R and I_L), shown as angle ϕ in the phasor diagram.

   From the phasor diagram:

   $I = \sqrt{(I_R^2 + I_L^2)}$, (by Pythagoras' theorem)

   where $I_R = \dfrac{V}{R}$ and $I_L = \dfrac{V}{X_L}$

   $\tan \phi = \dfrac{I_L}{I_R}$, $\sin \phi = \dfrac{I_L}{I}$ and $\cos \phi = \dfrac{I_R}{I}$

   (by trigonometric ratios)

   Circuit impedance, $Z = \dfrac{V}{I}$

3   **R-C parallel circuit**

   In the two branch parallel circuit containing resistance R and capacitance C shown in *Figure 14.1(b)*, I_R is in-phase with the supply voltage V and the current flowing in the capacitor, I_C, leads V by 90°. The supply current I is the phasor sum of I_R and I_C and thus the current I leads the applied voltage V by an angle lying between 0° and 90° (depending on the values of I_R and I_C), shown as angle α in the phasor diagram.

   From the phasor diagram:

   $I = \sqrt{(I_R^2 + I_C^2)}$, (by Pythagoras' theorem)

CIRCUIT DIAGRAM   PHASOR DIAGRAM

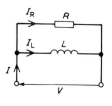
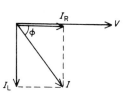

(a)

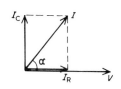

(b)

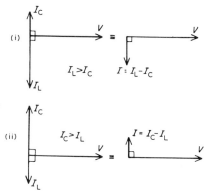

(c)

Figure 15.1

where $I_R = \dfrac{V}{R}$ and $I_C = \dfrac{V}{X_C}$

$\tan \alpha = \dfrac{I_C}{I_R}$, $\sin \alpha = \dfrac{I_C}{I}$ and $\cos \alpha = \dfrac{I_R}{I}$ (by trigonometric ratios)

Circuit impedance $Z = \dfrac{V}{I}$

97

4   *L-C* **parallel circuit**

In the two branch parallel circuit containing inductance L and capacitance C shown in *Figure 14.1(c)*, I_L lags V by 90° and I_C leads V by 90°. Theoretically there are three phasor diagrams possible — each depending on the relative values of I_L and I_C.

(i) $I_L > I_C$ (giving a supply current, $I = I_L - I_C$ lagging V by 90°)

(ii) $I_C > I_L$ (giving a supply current, $I = I_C - I_L$ leading V by 90°)

(iii) $I_L = I_C$ (giving a supply current, $I = 0$).

The latter condition is not possible in practice due to circuit resistance inevitably being present (as in the circuit described in para. 5).

For the *L-C* parallel circuit, $I_L = \dfrac{V}{X_L}$, $I_C = \dfrac{V}{X_C}$

$I =$ phasor difference between I_L and I_C, and $Z = \dfrac{V}{I}$.

5   *LR-C* **parallel circuit**

In the two branch circuit containing capacitance C in parallel with inductance L and resistance R in series (such as a coil) shown in *Figure 15.2(a)*, the phasor diagram for the *LR* branch alone is shown in *Figure 15.2(b)* and the phasor diagram for the C branch alone in *Figure 15.2(c)*. Rotating each and superimposing on one another gives the complete phasor diagram shown in *Figure 15.2(d)*.

6   The current I_{LR} of *Figure 15.2(d)* may be resolved into horizontal and vertical components. The horizontal component, shown as op is $I_{LR} \cos \phi_1$, and the vertical component, shown as pq is $I_{LR} \sin \phi_1$. There are three possible conditions for this circuit:

(i) $I_C > I_{LR} \sin \phi_1$ (giving a supply current I leading V by angle ϕ — as shown in *Figure 15.2(e)*).

(ii) $I_{LR} \sin \phi_1 > I_C$ (giving I lagging V by angle ϕ — as shown in *Figure 15.2(f)*).

(iii) $I_C = I_{LR} \sin \phi_1$ (this is called parallel resonance, see para. 10).

7   There are two methods of finding the phasor sum of currents I_{LR} and I_C in *Figure 15.2(e)* and *(f)*. These are: (i) by a scaled phasor diagram, or (ii) by resolving each current into their 'in-phase' (i.e. horizontal) and 'quadrature' (i.e. vertical) components.

8   With reference to the phasor diagrams of *Figure 15.2*:

Impedance of *LR* branch, $Z_{LR} = \sqrt{(R^2 + X_L^2)}$

Current $I_{LR} = \dfrac{V}{Z_{LR}}$ and $I_C = \dfrac{V}{X_C}$

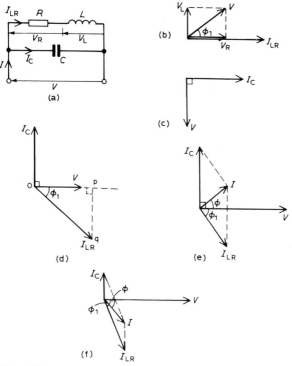

Figure 15.2

Supply current I = phasor sum of I_{LR} and I_C (by drawing)

$$= \sqrt{\{(I_{LR} \cos \phi_1) + (I_{LR} \sin \phi_1 \sim I_C)^2\}}$$

(by calculation)

where $\sim$ means 'the difference between'.

Circuit impedance $Z = \dfrac{V}{I}$

$\tan \phi_1 = \dfrac{V_L}{V_R} = \dfrac{X_L}{R}$, $\sin \phi_1 = \dfrac{X_L}{Z_{LR}}$ and $\cos \phi_1 = \dfrac{R}{Z_{LR}}$

$$\tan \phi = \frac{I_{LR} \sin \phi_1 \sim I_C}{I_{LR} \cos \phi_1} \text{ and } \cos \phi = \frac{I_{LR} \cos \phi_1}{I}$$

9   For any parallel a.c. circuit:

True or active power, $P = VI \cos \phi$ watts (W)
   or $P = I_R^2 R$ watts

Apparent power, $S = VI$ voltamperes (VA)

Reactive power, $Q = VI \sin \phi$ reactive voltamperes (VAr)

Power factor $= \dfrac{\text{true power}}{\text{apparent power}} = \dfrac{P}{S} = \cos \phi$.

(These formulae are the same as for series a.c. circuits.)

10   (i) **Resonance** occurs in the two branch circuit containing capacitance C in parallel with inductance L and resistance R in series (see *Figure 15.2(a)*) when the quadrature (i.e. vertical) component of current I_{LR} is equal to I_C. At this condition the supply current I is in-phase with the supply voltage V.

(ii) When the quadrature component of I_{LR} is equal to I_C then:

$$I_C = I_{LR} \sin \phi \text{ (see } Figure\ 15.3\text{)}$$

Hence $\dfrac{V}{X_C} = \left(\dfrac{V}{Z_{LR}}\right)\left(\dfrac{X_L}{Z_{LR}}\right)$, (from para. 8)

from which,

$$Z_{LR}^2 = X_C X_L = (2\pi f_r L)\left(\frac{1}{2\pi f_r C}\right) = \frac{L}{C} \qquad (1)$$

Hence $[\surd(R^2 + X_L^2)]^2 = \dfrac{L}{C}$ and $R^2 + X_L^2 = \dfrac{L}{C}$

$$(2\pi f_r L)^2 = \frac{L}{C} - R^2$$

$$f_r = \frac{1}{2\pi L} \sqrt{\left(\frac{L}{C} - R^2\right)} = \frac{1}{2\pi} \sqrt{\left(\frac{L}{L^2 C} - \frac{R^2}{L^2}\right)}$$

i.e. **parallel resonant frequency**, $f_r = \dfrac{1}{2\pi} \sqrt{\left(\dfrac{1}{LC} - \dfrac{R^2}{L^2}\right)}$ Hz

(When R is negligible, then $f_r = \dfrac{1}{2\pi\surd(LC)}$ which is the same as for series resonance.)

(iii) Current at resonance,

$$I_r = I_{LR} \cos \phi_1 \text{ (from Figure 15.3)}$$

$$= \left(\frac{V}{Z_{LR}}\right)\left(\frac{R}{Z_{LR}}\right) \text{ (from para. 8)}$$

$$= \frac{VR}{Z_{LR}^2}$$

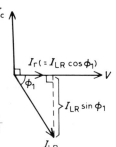

However, from equation (1), $Z_{LR}^2 = \dfrac{L}{C}$

Hence $I_r = \dfrac{VR}{\left(\dfrac{L}{C}\right)} = \dfrac{VRC}{L}$

Figure 15.3

The current is at a minimum at resonance.

(iv) Since the current at resonance is in-phase with the voltage the impedance of the circuit acts as a resistance. This resistance is known as the dynamic resistance, R_D (or sometimes, the dynamic impedance).

From equation (2), impedance at resonance $= \dfrac{V}{I_r} = \dfrac{L}{RC}$

i.e. **dynamic resistance**, $R_D = \dfrac{L}{RC} \ \Omega.$

(v) The parallel resonant circuit is often described as a **rejector** circuit since it presents its maximum impedance at the resonant frequency and the resultant current is a minimum.

11 Currents higher than the supply current can circulate within the parallel branches of a parallel resonant circuit, the current leaving the capacitor and establishing the magnetic field of the inductor, this then collapsing and recharging the capacitor, and so on. The **Q-factor** of a parallel resonant circuit is the ratio of the current circulating in the parallel branches of the circuit to the supply current, i.e. the current magnification.

Q-factor at resonance = current magnification

$$= \frac{\text{circulating current}}{\text{supply current}}$$

$$= \frac{I_C}{I_r} = \frac{I_{LR} \sin \phi_1}{I_r}$$

$$= \frac{I_{LR} \sin \phi_1}{I_{LR} \cos \phi_1} = \frac{\sin \phi_1}{\cos \phi_1}$$

$$= \tan \phi_1 = \frac{X_L}{R}$$

i.e. Q-factor at resonance $= \dfrac{2\pi f L}{R}$

(which is the same as for a series circuit).

Note that in a parallel circuit the Q-factor is a measure of current magnification, whereas in a series circuit it is a measure of voltage magnification.

12   At mains frequencies the Q-factor of a parallel circuit is usually low, typically less than 10, but in radio-frequency circuits the Q-factor can be very high.

13   For a particular power supplied, a high power factor reduces the current flowing in a supply system and therefore reduces the cost of cables, switch-gear, transformers and generators. Supply authorities use tariffs which encourage electricity consumers to operate at a reasonably high power factor.

Industrial loads such as a.c. motors are essentially inductive *(R-L)* and may have a low power factor. One method of improving (or correcting) the power factor of an inductive load is to connect a static capacitor C in parallel with the load (see *Figure 15.4(a)*). The supply current is reduced from I_{LR} to I, the phasor sum of I_{LR} and I_C, and the circuit power factor improves from $\cos \phi_1$ to $\cos \phi_2$ (see *Figure 15.4(b)*).

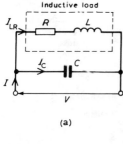

(a)

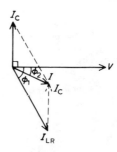

(b)

Figure 15.4

Another example where capacitors are connected directly across a load occurs in fluorescent lighting, where manufacturers include a power-factor correction capacitor inside the fitting.

14   Where a factory possesses a large number of a.c. motors it may be uneconomical to place a capacitor across the terminals of each motor. As the power factor of an individual motor may vary with load, the capacitor may result in overcorrection at certain loads and even produce a voltage surge that may have a damaging effect on the motor. Many factories have automatic power-factor correction plant situated in their substations, capacitors being switched in or out to maintain the system power-factor between certain predetermined limits.

16 a.c. circuit analysis

1   The laws which determine the currents and voltage drops in a.c. circuits are:

(a) current, $I = \dfrac{V}{Z}$

(b) the laws for impedances in series and parallel,
i.e., total impedance $Z_T = Z_1 + Z_2 + Z_3 + \cdots$
for impedances connected in series,

and $\dfrac{1}{Z_T} = \dfrac{1}{Z_1} + \dfrac{1}{Z_2} + \dfrac{1}{Z_3} + \cdots$

for impedances connected in parallel, and

(c) Kirchhoffs laws (as for d.c. circuits — see page 33).

2   A number such as $3 + j2$ is called a complex number, the 3 being the real part and $j2$ the imaginary part. By definition $j = \sqrt{-1}$ (*see Newnes Mathematics Pocket Book, page 81*).

Circuit theory and analysis of a.c. circuits is invariably achieved using complex numbers (otherwise known as symbolic or j notation).

The effect of multiplying a phasor by j is to rotate it in a positive direction (i.e. anticlockwise) on an Argand diagram through 90° without altering its length. Similarly, multiplying a phasor by $-j$ rotates the phasor through $-90°$. These facts are used in a.c. theory since certain quantities in the phasor diagrams lie at 90° to each other. For example, in the *R-L* series circuit shown in *Figure 16.1(a)*, V_L leads I by 90° (i.e. I lags V_L by 90°) and may be written as jV_L, the vertical axis being regarded as the imaginary axis of an Argand diagram. Thus $V_R + jV_L = V$ and since $V_R = IR$, $V_L = IX_L$ and $V = IZ$, then $R + jX_L = Z$.

Similarly, for the *R-C* circuit shown in *Figure 16.1(b)*, V_C lags I by 90° (i.e. I leads V_C by 90°) and $V_R - jV_C = V$, from which $R - jX_C = Z$. Thus an impedance of $(3 + j2)$ ohms represents a resistance of 3 Ω in series with an inductance of inductive reactance X_L of 2 Ω and an impedance of $(5 - j7)$ Ω represents a resistance of 5 Ω in series with a capacitance of capacitive reactance 7 Ω.

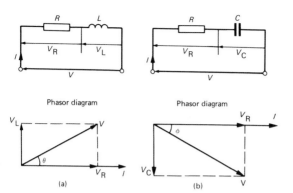

Figure 16.1

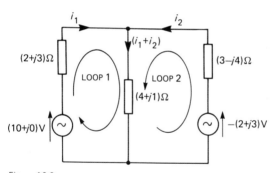

Figure 16.2

For example, to determine the current flowing in the $(4+j1)$ Ω impedance in *Figure 16.2*, currents are initially assigned to the three branches, as shown by i_1, i_2 and (i_1+i_2).

From loop 1:

$$10+j0 = i_1(2+j3) + (i_1+i_2)(4+j1)$$

i.e. $\qquad 10 = i_1(6+j4) + i_2(4+j1) \qquad (1)$

From loop 2:

$$-(2-j3) = i_2(3-j4) + (i_1+i_2)(4+j1)$$

i.e. $\qquad -(2+j3) = i_1(4+j1) + i_2(7-j3) \qquad (2)$

To eliminate i_2, equation (1) is multiplied by $(7-j3)$ and equation (2) by $(4+j1)$.

Thus,

$$10(7-j3) = i_1(6+j4)(7-j3) + i_2(4+j1)(7-j3) \quad (3)$$

$$-2(2+j3)(4+j1) = i_1(4+j)(4+j) + i_2(7-j3)(4+j) \quad (4)$$

Equation (3)–equation (4) gives:

$$10(7-j3) + (2+j3)(4+j) = i_1[(6+j4)(7-j3) - (4+j)(4+j)]$$

i.e.

$$70 - j30 + 8 + j2 + j12 - 3$$
$$= i_1[42 - j18 + j28 + 12 - 16 - j4 - j4 + 1]$$

$$75 - j16 = i_1(39+j2)$$

from which,

$$i_1 = \frac{75-j16}{39+j2} = \frac{76.69\angle -12.04°}{39.05\angle 2.94°} = \mathbf{1.96\angle -14.98° \ A \ or}$$
$$\mathbf{(1.89 - j0.51)A.}$$

From equation (1), $10 = (1.89 - j0.51)(6+j4) + i_2(4+j1)$

$$10 = 13.38 + j4.5 + i_2(4+j1)$$

i.e. $-3.38 - j4.5 = i_2(4+j1)$

$$i_2 = \frac{-3.38-j4.5}{4+j1} = \frac{(-3.38-j4.5)}{4^2+1^2}(4-j) = \mathbf{(-1.06 - j0.86)A}$$

Thus $(i_1 + u_2) = (1.89 - j0.51) + (-1.06 - j0.86) = \mathbf{(0.83 - j1.37)A \ or}$
$$\mathbf{1.60 \angle -59°A.}$$

3 There are a number of circuit theorems which have been developed for solving problems in a.c. electrical networks. These include:

    (a) the superposition theorem;
    (b) Thévénins theorem;
    (c) Norton's theorem, and
    (d) the maximum power transfer theorem.

As a preliminary to using circuit theorems, star-delta and delta-star transformations may be used.

4 **The superposition theorem** states:

'In any network made up of linear impedances and containing more than one source of emf, the resultant current flowing in any branch is the algebraic sum of the currents that would flow in that branch if

each source was considered separately, all other sources being replaced at that time by their respective internal impedances.'

For example, to determine the current flowing in each branch of the network shown in *Figure 16.2* using the superposition theorem, the procedure is:

(i) Redraw the original circuit with the $-(2+j3)$ volt source removed as shown in *Figure 16.3*.

(ii) Label the currents in each branch and their directions as shown in *Figure 16.3*, and determine their values.

$(4+j1)\,\Omega$ in parallel with $(3-j4)\,\Omega$ gives an equivalent impedance of

$$\frac{(4+j1)(3-j4)}{(4+j1)+(3-j4)} = \frac{16-j13}{7-j3} = \frac{(16-j13)(7+j3)}{7^2+3^2} = \frac{151-j43}{58}$$

$$= (2.60 - j0.74)\,\Omega$$

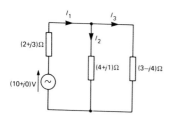

Figure 16.3

Hence from the equivalent circuit of *Figure 16.4*:

$$I_1 = \frac{10+j0}{(2+j3)+(2.60-j0.74)} = \frac{10}{4.60+j2.26} = \frac{10(4.60-j2.26)}{4.60^2+2.26^2}$$

i.e. $I_1 = \mathbf{(1.75 - j0.86)A}$.

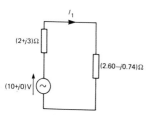

Figure 16.4

From *Figure 16.3*,

$$I_2 = \left[\frac{3 - j4}{(4 + j1) + (3 - j4)} \right](1.75 - j0.86)$$

$$= \left(\frac{3 - j4}{7 - j3} \right)(1.75 - j0.86) = \frac{1.81 - j9.59}{7 - j3}$$

$$= \frac{(1.81 - j9.58)(7 + j3)}{7^2 + 3^2} = \mathbf{(0.71 - j1.06)A}$$

and $I_3 = I_1 - I_2 = (1.75 - j0.86) - (0.71 - j1.06) = \mathbf{(1.04 + j0.20)A}$.
(iii) Redraw the original circuit with the $(10 + j0)$ volt source removed, as shown in *Figure 16.5*.

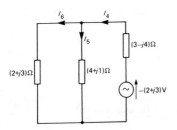

Figure 16.5

(iv) Label the currents in each branch and their directions as shown in *Figure 16.5* and determine their values.
$(2 + j3)\,\Omega$ in parallel with $(4 + j1)\,\Omega$ gives an equivalent impedance of

$$\frac{(2 + j3)(4 + j1)}{(2 + j3) + (4 + j1)} = \frac{5 + j14}{6 + j4} = \frac{(5 + j14)(6 - j4)}{(6 + j4)(6 - j4)} = \frac{76 + j64}{52}$$

$$= (1.46 + j1.23)\,\Omega$$

From the equivalent circuit of *Figure 16.6*

$$I_4 = \frac{-(2 + j3)}{(4.46 - j2.77)} = \frac{(-2 - j3)(4.46 + j2.77)}{4.46^2 + 2.77^2} = \frac{-0.61 - j18.92}{27.56}$$

$$= \mathbf{(-0.022 - j0.687)A}$$

From *Figure 16.5*,

$$I_5 = \left[\frac{2 + j3}{(2 + j3)(4 + j1)} \right](-0.022 - j0.687)$$

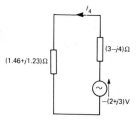

Figure 16.6

$$= \frac{(2+j3)}{(6+j4)}(-0.022-j0.687) = \frac{2.017-j1.44}{6+j4}$$

$$= \frac{(2.017-j1.44)(6-j4)}{6^2+4^2} = \frac{6.342-j16.708}{52}$$

$$= (0.122 - j0.321)\text{A}$$

From *Figure 16.5*,

$$I_6 = I_4 - I_5 = (-0.022 - j0.687) - (0.122 - j0.321)$$
$$= (-0.144 - j0.366)\text{A}.$$

v) Superimposing *Figure 16.3* on to *Figure 16.5* gives:

$$(I_1 - I_6) = (1.75 - j0.86) - (-0.144 - j0.366)$$
$$= (1.89 - j0.49)\text{A or } 1.96 \angle -14.62° \text{ A}$$
$$(I_4 - I_3) = (-0.022 - j0.687) - (1.04 + j0.20)$$
$$= (-1.06 - j0.89)\text{A or } 1.38 \angle -140°$$

and

$$(I_2 + I_5) = (0.71 - j1.06) + (0.122 - j0.321)$$
$$= (0.83 - j1.38)\text{A or } 1.61 \angle -59°$$

as shown in *Figure 16.7*.

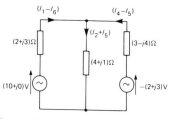

Figure 16.7

5   (a) **Thévenin's theorem** states:

'*The current in any branch of a network is that which would result if an emf, equal to the p.d. across a break made in the branch, were introduced into the branch, all other emf's being removed and represented by the internal impedances of the sources.*'

(b) To determine the current in any branch of an active network the procedure adopted is:

(i) remove the impedance Z from that branch;

(ii) determine the open-circuit voltage E across the break;

(iii) remove each source of emf and replace them by their internal impedances and then determine the impedance z 'looking-in' at the break,

(iv) determine the value of the current from the equivalent circuit shown in *Figure 16.8*.

i.e. $i = \dfrac{E}{Z + z}$

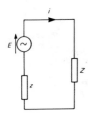

Figure 16.8

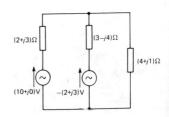

Figure 16.9

For example, to determine the current flowing in the $(4+j1)\,\Omega$ impedance shown in *Figure 16.9* using Thévenin's theorem (note *Figure 16.9* is the same as *Figure 16.2*):

(i) The impedance $(4+j1)\,\Omega$ is removed as shown in *Figure 16.10*.

(ii) Current $I_1 = \dfrac{(10+j0) - (-2-j3)}{(2+j3) + (3-j4)} = \dfrac{12+j3}{5-j1} = \dfrac{(12+j3)(5+j1)}{(5-j1)(5+j1)}$

$$= \frac{57+j27}{26} = (2.19 + j1.04)\,\text{A}$$

p.d. across AB,

$E = (10 + j0) - (2.19 + j1.04)(2 + j3)$
$= 10 - (1.26 + j8.65) = (8.74 - j8.65)$ volts.

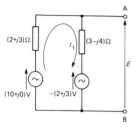

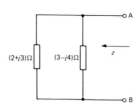

Figure 16.10   Figure 16.11

(iii) Removing the sources of emf gives the circuit of *Figure 16.11*, from which,

$$\text{impedance } z = \frac{(2+j3)(3-j4)}{(2+j3)+(3-j4)} = \frac{(18+j1)}{(5-j1)}$$

$$= \frac{(18+j1)(5+j1)}{(5-j1)(5+j1)}$$

$$= \frac{89+j23}{26} = (3.42 - j0.88)\ \Omega$$

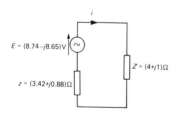

Figure 16.12

(iv) The equivalent Thévénin circuit is shown in *Figure 16.12*.

$$\text{Current } i = \frac{8.74 - j8.65}{(3.42 + j0.88) + (4 + j1)} = \frac{8.74 - j8.65}{7.42 + j1.88}$$

$$= \frac{12.30\angle -44.78°}{7.65\angle 14.22°}$$

i.e. current in $(4+j1)\ \Omega$ impedance, $i = \mathbf{1.61}\angle -\mathbf{59°A}$.

111

6   (a) **Norton's theorem** states:

> '*The current that flows in any branch of a network is the same as that which would flow in the branch if it were connected across a source of electricity, the short-circuit current of which is equal to the current that would flow in a short-circuit across the branch, and the internal impedance of which is equal to the impedance which appears across the open-circuited branch terminals.*'

(b) To determine the current in any branch AB of an active network the procedure adopted is:

  (i) short-circuit that branch;

  (ii) determine the short-circuit current, I_{SC};

  (iii) remove each source of emf and replace them by their internal impedances (or, if a current source exists replace with an open-circuit), then determine the impedance, z, 'looking-in' at a break made between A and B,

  (iv) determine the value of the current from the equivalent circuit shown in *Figure 16.13*,

i.e. $i = \left(\dfrac{z}{Z+z}\right) I_{SC}$

For example, to determine the current flowing in the $(4+j1)\,\Omega$ impedance shown in *Figure 16.9*, using Norton's theorem:

(i) The branch containing the impedance $(4+j1)\,\Omega$ is short-circuited as shown in *Figure 16.14*

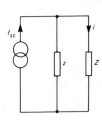

Figure 16.13

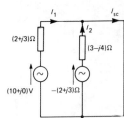

Figure 16.14

(ii) The short-circuit current

$I_{SC} = I_1 + I_2$

$= \dfrac{10}{(2+j3)} + \dfrac{-(2+j3)}{(3-j4)}$

$= (1.538 - j2.308) + (0.24 - j0.68)$

$= (1.78 - j2.99)\,\text{A}$

(iii) Impedance z 'looking-in' at break in branch,
$z = (3.42 + j0.88)$ from (iii) of para. 5.
(iv) From the equivalent circuit of *Figure 16.13*,

$$i = \left(\frac{z}{Z + z}\right) I_{SC}$$

i.e.

$$I = \left[\frac{(3.42 + j0.88)}{(4 + j1) + (3.42 + j0.88)}\right](1.78 - j2.99)$$

$$= \frac{(3.42 + j0.88)(1.78 - j2.99)}{7.42 + j1.88}$$

$$= \frac{(3.53 \angle 14.43°)(3.48 \angle -59.23°)}{7.65 \angle 14.22°}$$

Hence current flowing in $(4 + j1)\ \Omega$ impedance, $i = \mathbf{1.61} \angle \mathbf{-59°}$

7 Maximum power transfer theorem

Any network containing one or more current or voltage sources
and linear impedances can be reduced to a Thévenin equivalent
circuit. When a load is connected to the terminals of this
equivalent circuit, power is transferred from the circuit to the load.

A general Thévenin equivalent circuit is shown in *Figure 16.15*
with source internal impedance $(r + jx)\ \Omega$ and complex load
$(R + jX)\ \Omega$.

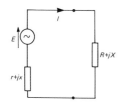

Figure 16.15

(i) For a purely resistive load and source internal impedance,
when $X = x = 0$, maximum power is transferred when $R = r$
(ii) When the load resistance R and reactance X are both
independantly adjustable, maximum power transfer is obtained
when $X = -x$ **and** $R = r$.
(iii) When the load is purely resistive and adjustable (i.e. $X = 0$),
maximum power transfer is obtained when $R = |z|$, where
$|z| = \sqrt{(r^2 + x^2)}$.

(iv) When the load resistance R is adjustable but reactance X is fixed, maximum power transfer is obtained when
$R = \sqrt{[r^2 + (X + x)^2]}$.

8   Sometimes networks are complicated and may be transformed using delta-star or star-delta transformations as a preliminary to using a circuit theorem.

(a) *Delta-star conversion*

Given impedances Z_P, Z_Q and Z_R connected in delta as shown in *Figure 16.16(a)*, then the equivalent in star connection (see *Figure 16.16(b)*) is given by:

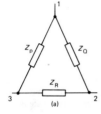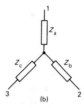

Figure 16.16

$$Z_a = \frac{Z_P Z_Q}{Z_P + Z_Q + Z_R}, \ Z_b = \frac{Z_Q Z_R}{Z_P + Z_Q + Z_R} \text{ and } Z_c = \frac{Z_R Z_P}{Z_P + Z_Q + Z_R}$$

(b) *Star-delta conversion*

Given impedances Z_a, Z_b and Z_c connected in star, as shown in *Figure 16.16(b)*, then the equivalent in delta connection (*Figure 16.16(a)*) is given by:

$$Z_P = \frac{Z_a Z_b + Z_b Z_c + Z_c Z_a}{Z_b}, \ Z_Q = \frac{Z_a Z_b + Z_b Z_c + Z_c Z_a}{Z_c}$$
$$\text{and } Z_R = \frac{Z_a Z_b + Z_b Z_c + Z_c Z_a}{Z_a}$$

17 Semiconductor diodes

1   Materials may be classified as conductors, semiconductors or insulators. The classification depends on the value of resistivity of the material. Good conductors are usually metals and have resistivities in the order of 10^{-7} to 10^{-8} ohm metres. Semiconductors have resistivities in the order of 10^{-3} to 3×10^{3} Ωm. The resistivities of insulators are in the order of 10^{4} to 10^{14} Ω. Some typical approximate values at normal room temperatures are:

Conductors:	Aluminium	$2.7 \times 10^{-8}\,\Omega$m
	Brass (70 Cu/30 Zn)	$8 \times 10^{-8}\,\Omega$m
	Copper (pure annealed)	$1.7 \times 10^{-8}\,\Omega$m
	Steel (mild)	$15 \times 10^{-8}\,\Omega$m
Semiconductors:	Silicon $\quad$ $2.3 \times 10^{3}\,\Omega$m $\,\big\}$	at 27°C
	Germanium $\quad 0.45\,\Omega$m	
Insulators:	Glass $\geqslant 10^{10}\,\Omega$m	
	Mica $\geqslant 10^{11}\,\Omega$m	
	P.V.C. $\geqslant 10^{13}\,\Omega$m	
	Rubber (pure) 10^{12} to $10^{14}\,\Omega$m.	

2   In general, over a limited range of temperatures, the resistance of a conductor increases with temperature increase. The resistance of insulators remains approximately constant with variation of temperature. The resistance of semiconductor materials decreases as the temperature increases. For a specimen of each of these materials, having the same resistance (and thus completely different dimensions), at, say, 15°C, the variation for a small increase in temperature to t°C is as shown in *Figure 17.1*.

3   The most important semiconductors used in the electronics industry are **silicon** and **germanium**. As the temperature of these materials is raised above room temperature, the resistivity is reduced and ultimately a point is reached where they effectively become conductors. For this reason, silicon should not operate at a working temperature in excess of 150°C to 200°C, depending on its purity, and germanium should not operate at a working temperature in excess of 75°C to 90°C, depending on its purity. As the temperature of a semiconductor is reduced below normal room

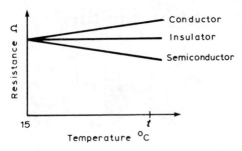

Figure 17.1

temperature, the resistivity increases until, at very low
temperatures, the semiconductor becomes an insulator.

4   Adding extremely small amounts of impurities to pure
semiconductors in a controlled manner is called **doping**.
Antimony, arsenic and phosphorus are called n-type impurities and
form an **n-type material** when any of these impurities are added
to silicon or germanium. The amount of impurity added usually
varies from 1 part impurity in 10^5 parts semiconductor material to
1 part impurity to 10^8 parts semiconductor material, depending on
the resistivity required. Indium, aluminium and boron are called p-
type impurities and form a **p-type material** when any of these
impurities are added to a semiconductor.

5   In semiconductor materials, there are very few charge carriers
per unit volume free to conduct. This is because the 'four electron
structure' in the outer shell of the atoms (called valency electrons),
form strong covalent bonds with neighbouring atoms, resulting in a
tetrahedral structure with the electrons held fairly rigidly in place.
A two-dimensional diagram depicting this is shown for germanium
in *Figure 17.2*.

Arsenic, antimony and phosphorus have five valency electrons
and when a semiconductor is doped with one of these substances,
some impurity atoms are incorporated in the tetrahedral structure.
The 'fifth' valency electron is not rigidly bonded and is free to
conduct, the impurity atom donating a charge carrier. A two-
dimensional diagram depicting this is shown in *Figure 17.3*, in
which a phosphorus atom has replaced one of the germanium
atoms. The resulting material is called n-type material, and
contains free electrons.

Indium, aluminium and boron have three valency electrons
and when a semiconductor is doped with one of these substances,

116

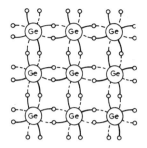

Figure 17.2

Figure 17.3

some of the semiconductor atoms are replaced by impurity atoms. One of the four bonds associated with the semiconductor material is deficient by one electron and this deficiency is called a **hole**. Holes give rise to conduction when a potential difference exists across the semiconductor material due to movement of electrons from one hole to another, as shown in *Figure 17.4*. In this figure, an electron moves from A to B, giving the appearance that the hole moves from B to A. Then electron C moves to A, giving the appearance that the hole moves to C, and so on. The resulting material is *p*-type material containing holes.

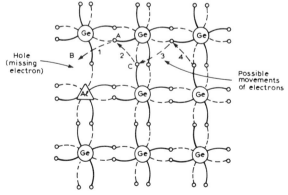

Figure 17.4

6   A **p-n junction** is a piece of semiconductor material in which part of the material is *p*-type and part is *n*-type. In order to examine the charge situation, assume that separate blocks of *p*-type and *n*-type materials are pushed together. Also assume that a hole is a positive charge carrier and that an electron is a negative charge carrier. At the junction, the donated electrons in the *n*-type material, called **majority carriers**, diffuse into the *p*-type material (diffusion is from an area of high density to an area of lower density) and the acceptor holes in the *p*-type material diffuse into the *n*-type material as shown by the arrows in *Figure 17.5*.

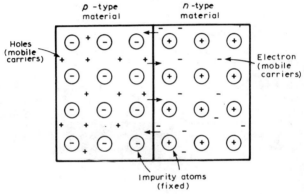

Figure 17.5

Because the *n*-type material has lost electrons, it acquires a positive potential with respect to the *p*-type material and thus tends to prevent further movement of electrons. The *p*-type material has gained electrons and becomes negatively charged with respect to the *n*-type material and hence tends to retain holes. Thus after a short while, the movement of electrons and holes stops due to the potential difference across the junction, called the **contact potential**. The area in the region of the junction becomes depleted of holes and electrons due to electron-hole recombinations, and is called a **depletion layer**, as shown in *Figure 17.6*.

7   When an external voltage is applied to a *p-n* junction making the *p*-type material positive with respect to the *n*-type material, as shown in *Figure 17.7*, the *p-n* junction is **forward biased**. The applied voltage opposes the contact potential, and, in effect, closes

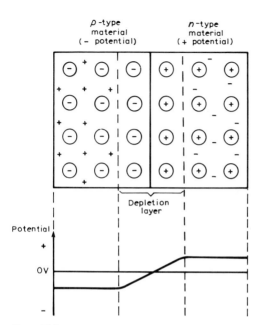

Figure 17.6

the depletion layer. Holes and electrons can now cross the junction and a current flows. An increase in the applied voltage above that required to narrow the depletion layer (about 0.2 V for germanium and 0.6 V for silicon), results in a rapid rise in the current flow. Graphs depicting the current-voltage relationship for forward biased *p-n* junctions, for both germanium and silicon, called the forward characteristics, are shown in *Figure 17.8*.

When an external voltage is applied to a *p-n* junction making the *p*-type material negative with respect to the *n*-type material, as shown in *Figure 17.9*, the *p-n* junction is **reverse biased**. The applied voltage is now in the same sense as the contact potential and opposes the movement of holes and electrons, due to opening up the depletion layer. Thus, in theory, no current flows. However, at normal room temperature, certain electrons in the covalent bond lattice acquire sufficient energy from the heat available to leave the lattice, generating mobile electrons and holes. This process is called electron-hole generation by thermal excitation.

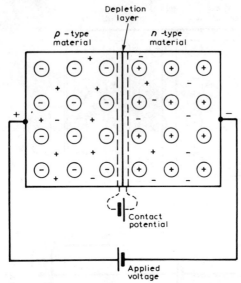

Figure 17.7

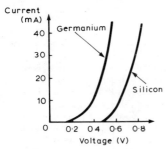

Figure 17.8

The electrons in the *p*-type material and holes in the *n*-type material caused by thermal excitation, are called **minority carriers** and these will be attracted by the applied voltage. Thus, in practice, a small current of a few micro-amperes for germanium and less than one micro-ampere for silicon, at normal room

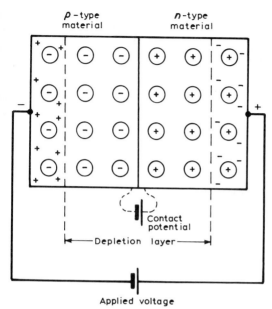

Figure 17.9

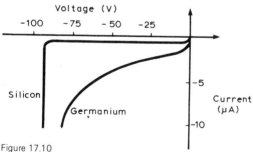

Figure 17.10

temperature, flows under reverse bias conditions. Typical reverse characteristics are shown in *Figure 17.10* for both germanium and silicon.

8   A **semiconductor diode** is a device having a *p-n* junction mounted in a container, suitable for conducting and dissipating the

 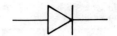

BSI recommended
circuit symbol

IEE recommended
circuit symbol

Figure 17.11

heat generated in operation and having connecting leads. Its operating characteristics are as shown in *Figures 17.8* and *17.10*. Two circuit symbols for semiconductor diodes are in common use and are shown in *Figure 17.11*.

18 Transistors

1   The bipolar junction transistor consists of three regions of semiconductor material. One type is called a *p-n-p* transistor, in which two regions of *p*-type material sandwich a very thin layer of *n*-type material. A second type is called a *n-p-n* transistor, in which two regions of *n*-type material sandwich a very thin layer of *p*-type material. Both of these types of transistors consist of two *p-n* junctions placed very close to one another in a back-to-back arrangement on a single piece of semiconductor material. Diagrams depicting these two types of transistor are shown in *Figure 18.1*.

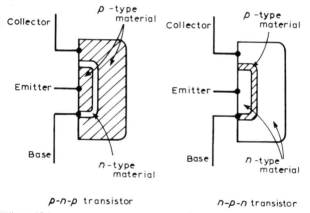

p-n-p transistor   *n-p-n* transistor

Figure 18.1

   The two *p*-type material regions of the *p-n-p* transistor are called the emitter and collector and the *n*-type material is called the base. Similarly, the two *n*-type material regions of the *n-p-n* transistor are called the emitter and collector and the *p*-type material region is called the base, as shown in *Figure 18.1*.
   Transistors have three connecting leads and in operation an

123

electrical input to one pair of connections, say the emitter and base
connections can control the output from another pair, say the
collector and emitter connections. This type of operation is
achieved by appropriately biasing the two internal *p-n* junctions.
When batteries and resistors are connected to a *p-n-p* transistor, as
shown in *Figure 18.2(a)*, the base-emitter junction is forward biased
and the base-collector junction is reverse biased. Similarly, an *n-p-n*
transistor has its base-emitter junction forward biased and its base-
collector junction reverse biased when the batteries are connected
as shown in *Figure 18.2(b)*.

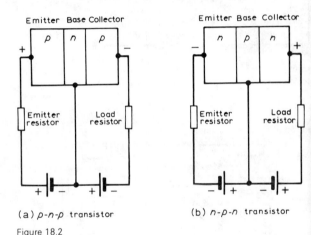

(a) *p-n-p* transistor

(b) *n-p-n* transistor

Figure 18.2

3   For a silicon *p-n-p* transistor, biased as shown in *Figure
18.2(a)*, if the base-emitter junction is considered on its own, it is
forward biased and a current flows. This is depicted in *Figure
18.3(a)*. For example, if R_E is 1000 Ω, the battery is 4.5 V and the
voltage drop across the junction is taken as 0.7 V, the current
flowing is given by $(4.5-0.7)/1000=3.8$ mA. When the base-
collector junction is considered on its own, as shown in *Figure
18.3(b)*, it is reverse biased and the collector current is something
less than one microampere.

However, when both external circuits are connected to the
transistor, most of the 3.8 mA of current flowing in the emitter,
which previously flowed from the base connection, now flows out
through the collector connection due to transistor action.

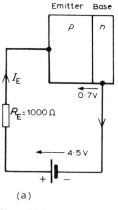

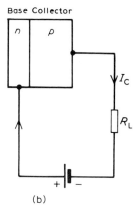

I_E

$R_E = 1000\,\Omega$

0·7V

4·5 V

I_C

R_L

Emitter   Base

Base   Collector

(a)   (b)

Figure 18.3

4   In a *p-n-p* transistor, connected as shown in *Figure 18.2(a)*, transistor action is accounted for as follows:

  (a)The majority carriers in the emitter *p*-type material are holes.

  (b)The base-emitter junction is forward biased to the majority carriers and the holes cross the junction and appear in the base region.

  (c)The base region is very thin and is only lightly doped with electrons so although some electron-hole pairs are formed, many holes are left in the base region.

  (d)The base-collector junction is reverse biased to electrons in the base region and holes in the collector region, but forward biased to holes in the base region. These holes are attracted by the negative potential at the collector terminal.

  (e)A large proportion of the holes in the base region cross the base-collector junction into the collector region, creating a collector current. Conventional current flow is in the direction of hole movement.

The transistor action is shown diagrammatically in *Figure 18.4*.

For transistors having very thin base regions, up to $99\frac{1}{2}\%$ of the holes leaving the emitter cross the base collector junction.

  In an *n-p-n* transistor, connected as shown in *Figure 18.2(b)*, transistor action is accounted for as follows:

125

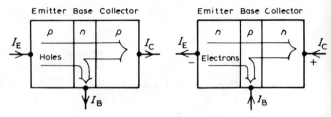

Figure 18.4   Figure 18.5

(a) The majority carriers in the *n*-type emitter material are electrons.
(b) The base-emitter junction is forward biased to these majority carriers and electrons cross the junction and appear in the base region.
(c) The base region is very thin and only lightly doped with holes, so some recombination with holes occurs but many electrons are left in the base region.
(d) The base-collector junction is reverse biased to holes in the base region and electrons in the collector region, but is forward biased to electrons in the base region. These electrons are attracted by the positive potential at the collector terminal.
(e) A large proportion of the electrons in the base region cross the base-collector junction into the collector region, creating a collector current.

The transistor action is shown diagrammatically in *Figure 18.5*. As stated in para. 4, conventional current flow is taken to be in the direction of hole flow, that is, in the opposite direction to electron flow, hence the directions of the conventional current flow are as shown in *Figure 18.5*.

6   For a *p-n-p* transistor, the base-collector junction is reverse biased for majority carriers. However, a small leakage current, I_{CBo} flows from the base to the collector due to thermally generated minority carriers (electrons in the collector and holes in the base), being present.

The base-collector junction is forward biased to these minority carriers. If a proportion, α, (having a value of up to 0.99 in modern transistors), of the holes passing into the base from the emitter, pass through the base-collector junction, then the various currents flowing in a *p-n-p* transistor are as shown in *Figure 18.6*.

Similarly, for a *n-p-n* transistor, the base-collector junction is reverse biased for majority carriers, but a small leakage current, I_{CBO}, flows from the collector to the base due to thermally

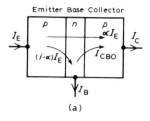

(a)

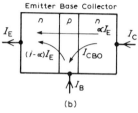

(b)

Figure 18.6

generated minority carriers (holes in the collector and electrons in the base), being present. The base-collector junction is forward biased to these minority carriers. If a proportion, α, of the electrons passing through the base-emitter junction also pass through the base-collector junction, then the currents flowing in an *n-p-n* transistor are as shown in *Figure 18.6*.

7   Symbols are used to represent *p-n-p* and *n-p-n* transistors in circuit diagrams and two of these in common use are shown in *Figure 18.7*. The arrow head drawn on the emitter of the symbol is in the direction of conventional emitter current (hole flow). The potentials marked at the collector, base and emitter are typical values for a silicon transistor having a potential difference of 6 V between its collector and its emitter.

The voltage of 0.6 V across the base and emitter is that required to reduce the potential barrier and if it is raised slightly to, say, 0.62 V, it is likely that the collector current will double to about 2 mA. Thus a small change of voltage between the emitter and the base can give a relatively large change of current in the emitter circuit. Because of this, transistors can be used as amplifiers.

There are three ways of connecting a transistor, depending on the use to which it is being put. The ways are classified by the electrode which is common to both the input and the output.

127

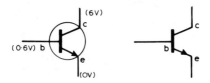

p-n-p transistor

Figure 18.7 n-p-n transistor

They are called

(a) common-base configuration, shown in *Figure 18.8(a)*,

(b) common-emitter configuration, shown in *Figure 18.8(b)*, and

(c) common-collector configuration, shown in *Figure 18.8(c)*.

These configurations are for an n-p-n transistor. The current flows shown are all reversed for a p-n-p transistor.

9   The effect of changing one or more of the various voltages and currents associated with a transistor circuit can be shown graphically and these graphs are called the characteristics of the transistor. As there are five variables (collector, base and emitter currents and voltages across the collector and base and emitter and base) and also three configurations, many characteristics are possible. Some of the possible characteristics are given below.

(a) Common-base configuration

(i) **Input characteristic**. With reference to *Figure 18.8(a)*, the input to a common-base transistor is the emitter current, I_E, and can be varied by altering the base-emitter voltage V_{EB}. The base-emitter junction is essentially a forward biased junction diode, so as V_{EB} is varied, the current flowing is similar to that for a junction diode, as

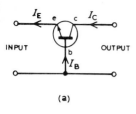

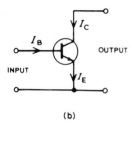

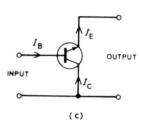

Figure 18.8

shown in *Figure 18.9* for a silicon transistor. *Figure 18.9* is called the input characteristic for an *n-p-n* transistor having common-base configuration. The variation of the collector-base voltage V_{CB} has little effect on the characteristic. A similar characteristic can be obtained for a *p-n-p* transistor, these having reversed polarities.

(ii) **Output characteristics** The value of the collector current I_C is very largely determined by the emitter current, I_E. For a given value of I_E, the collector-base voltage, V_{CB}, can be varied and has little effect on the value of I_C. If V_{CB} is made slightly negative, the collector no longer attracts the majority carriers leaving the emitter and I_C falls rapidly to zero. A family of curves for various values of I_E are possible and some of these are shown in *Figure 18.10*. *Figure 18.10* is called the output characteristic from an *n-p-n* transistor having common-base configuration. Similar characteristics can be obtained for a *p-n-p* transistor, these having reversed polarities.

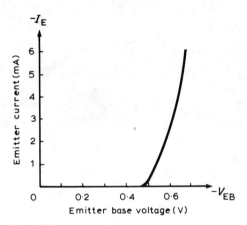

Figure 18.9

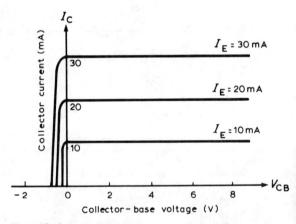

Figure 18.10

(b) Common-emitter configuration

(i) **Input characteristic** In a common-emitter configuration (see *Figure 18.8(b)*), the base current is now the input current. As V_{EB} is varied, the characteristic obtained is similar in shape to the input characteristic for a

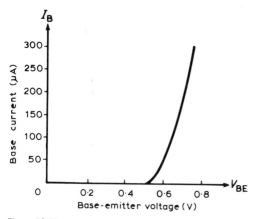

Figure 18.11

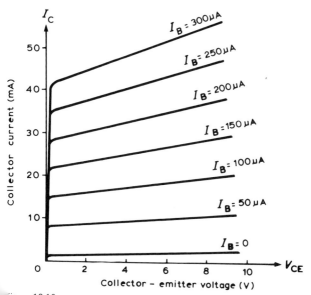

Figure 18.12

131

common-base configuration shown in *Figure 18.9*, but the values of current are far less. With reference to *Figure 18.6(a)*, as long as the junctions are biased as described, the three currents, I_E, I_C and I_B keep the ratio $1 : \alpha : (1 - \alpha)$ whichever configuration is adopted.

Thus the base current changes are much smaller than the corresponding emitter current changes and the input characteristic for an *n-p-n* transistor is as shown in *Figure 18.11*. A similar characteristic can be obtained for a *p-n-p* transistor, these having reversed polarities.

(ii) **Output characteristics** A family of curves can be obtained depending on the value of base current I_B and some of these for an *n-p-n* transistor are shown in *Figure 18.12*. A similar set of characteristics can be obtained for a *p-n-p* transistor, these having reversed polarities. These characteristics differ from the common-base output characteristics in the following ways:

The collector current reduces to zero without having to reverse the collector voltage, and

The characteristics slope upwards indicating a lower output resistance (usually kilohms for a common-emitter configuration compared with megohms for a common-base configuration).

19 Three-phase systems

1   Generation, transmission and distribution of electricity via the National Grid system is accomplished by three-phase alternating currents.

2   The voltage induced by a single coil when rotated in a uniform magnetic field is shown in *Figure 19.1* and is known as a **single-phase voltage**. Most consumers are fed by means of a single-phase a.c. supply. Two wires are used, one called the live conductor (usually coloured red) and the other is called the neutral conductor (usually coloured black). The neutral is usually connected via protective gear to earth, the earth wire being coloured green. The standard voltage for a single-phase a.c. supply

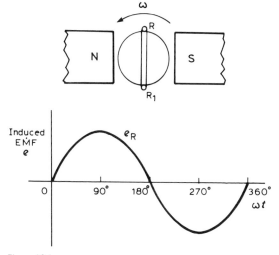

Figure 19.1

is 240 V. The majority of single-phase supplies are obtained by connection to a three-phase supply (see *Figure 19.5*).

3   **A three-phase supply** is generated when three coils are placed 120° apart and the whole rotated in a uniform magnetic field as shown in *Figure 19.2(a)*. The result is three independent supplies of equal voltages which are each displaced by 120° from each other as shown in *Figure 19.2(b)*.

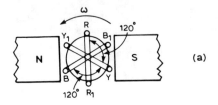

(a)

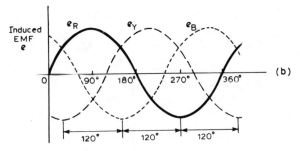

(b)

Figure 19.2

4   (i) The convention adopted to identify each of the phase voltages is: R—Red, Y—Yellow and B—blue, as shown in *Figure 19.2*.
(ii) The **phase-sequence** is given by the sequence in which the conductors pass the point initially taken by the red conductor. The national standard phase sequence is R, Y, B.

5   A three-phase a.c. supply is carried by three conductors, called '**lines**' which are coloured red, yellow and blue. The currents in these conductors are known as line currents (I_L) and the p.d.'s between them are known as line voltages (V_L). A fourth conductor, called the neutral (coloured black, and connected

134

through protective devices to earth) is often used with a three-phase supply.

6   If the three-phase windings shown in *Figure 19,2* are kept independent then six wires are needed to connect a supply source (such as a generator) to a load (such as a motor). To reduce the number of wires it is usual to interconnect the three phases. There are two ways in which this can be done, these being: (a) a star connection, and (b) a delta, or mesh, connection. Sources of three-phase supplies, i.e. alternators, are usually connected in star, whereas three-phase transformer windings, motors and other loads may be connected either in star or delta.

7   (i) **A star-connected load** is shown in *Figure 19.3* where the three line conductors are each connected to a load and the outlets from the loads are joined together at N to form what is termed the **neutral point** or the **star point**.

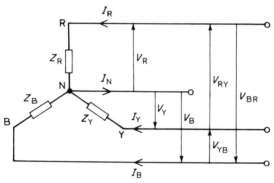

Figure 19.3

(ii) The voltages, V_R, V_Y and V_B are called **phase voltages** or line to neutral voltages. Phase voltages are generally denoted by V_p.
(iii) The voltages, V_{RY}, V_{YB} and V_{BR} are called **line voltages**.
(iv) From *Figure 19.3* it can be seen that the phase currents (generally denoted by I_p) are equal to their respective line currents I_R, I_Y and I_B, i.e. for a star connection:

$$\boxed{I_L = I_p}$$

(v) For a balanced system: $I_R = I_Y = I_B$, $V_R = V_Y = V_B$, $V_{RY} = V_{YB} = V_{BR}$, $Z_R = Z_Y = Z_B$ and the current in the neutral conductor, $I_N = 0$. When a star connected system is balanced, then the neutral conductor is unnecessary and is often omitted.

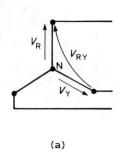

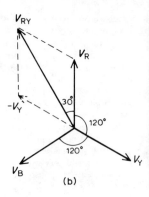

(a)

(b)

Figure 19.4

(vi) The line voltage, V_{RY}, shown in *Figure 19.4(a)* is given by $V_{RY} = V_R - V_Y$. (V_Y is negative since it is in the opposite direction to V_{RY}.) In the phasor diagram of *Figure 19.4(b)*, phasor V_Y is reversed (shown by the broken line) and then added phasorially to V_R, (i.e. $V_{RY} = V_R + (-V_Y)$). By trigonometry, or by measurement, $V_{RY} = \sqrt{3}V_R$, i.e. for a balanced star connection:

$$\boxed{V_L = \sqrt{3}V_p}$$

A phasor diagram for a balanced, three-wire, star-connected, 3-phase load having a phase voltage of 240 V, a line current of 5 A and a lagging power factor of 0.966 is shown in *Figure 19.5*. The phasor diagram is constructed as follows:

(i) Draw $V_R = V_Y = V_B = 240$ v and spaced 120° apart.
(ii) Power factor $= \cos \phi = 0.966$ lagging. Hence the load phase angle is given by arccos 0.966, i.e. 15° lagging. Hence

$$I_R = I_Y = I_B = 5\text{A}$$

lagging V_R, V_Y and V_B respectively by 15°.
(iii) $V_{RY} = V_R - V_Y$ (phasorically). By measurement, $V_{RY} = 415$ V (i.e. $\sqrt{3}(240)$) and leads V_R by 30°. Similarly, $V_{YB} = V_Y - V_B$ and $V_{BR} = V_B - V_R$.

(vii) The star connection of the three phases of a supply, together with a neutral conductor, allows the use of two voltages — the phase voltage and the line voltage. A 4-wire system is also used when the load is not balanced. The standard electricity supply to consumers in Great Britain is **415/240 V, 50 Hz, 3-phase, 4-wire alternating voltage**, and a diagram of connections is shown in *Figure 19.6*.

8 (i) **A delta (or mesh) connected load** is shown in *Figure*

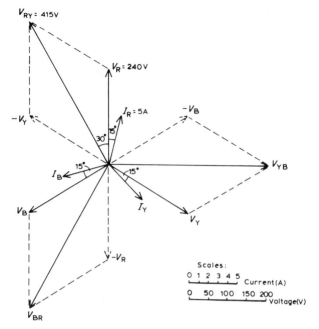

Figure 19.5

19.7 where the end of one load is connected to the start of the next load.

(ii) From *Figure 19.7*, it can be seen that the line voltages V_{RY}, V_{YB} and V_{BR} are the respective phase coltages, i.e. for a delta connection:

$$\boxed{V_L = V_p}$$

(iii) Using Kirchhoff's current law in *Figure 19.7*,

$$I_R = I_{RY} - I_{BR} = I_{RY} + (-I_{BR}).$$

From the phasor diagram shown in *Figure 19.8*, by trigonometry or by measurement, $I_R = \sqrt{3}\,I_{RY}$, i.e. for a delta connection:

$$\boxed{I_L = \sqrt{3}\,I_p}$$

9   The power dissipated in a three-phase load is given by the

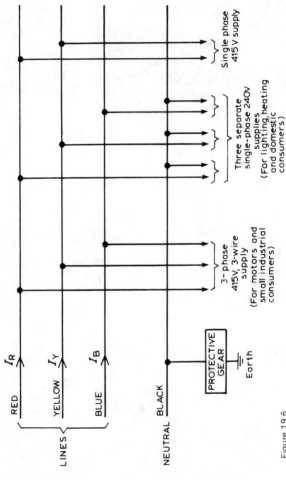

Figure 19.6

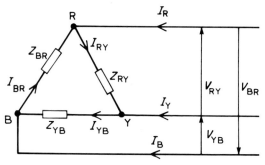

Figure 19.7

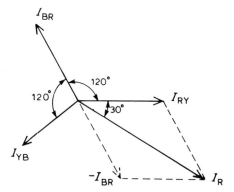

Figure 19.8

sum of the power dissipated in each phase. If a load is balanced then the total power P is given by:

$P = 3 \times$ power consumed by one phase.

The power consumed in one phase $= I_P^2 R_P$ or $V_P I_P \cos \phi$ (where ϕ is the phase angle between V_p and I_p).

For a star connection

$V_p = V_L/\sqrt{3}$ and $I_p = I_L$

hence

$P = 3(V_L/\sqrt{3})I_L \cos \phi = \sqrt{3} V_L I_L \cos \phi.$

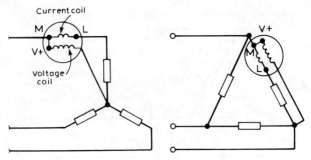

Figure 19.9

For a delta connection

$V_P = V_L$ and $I_p = I_L/\sqrt{3}$

hence

$P = 3V_L(I_L/\sqrt{3}) \cos \phi = \sqrt{3}V_LI_L \cos \phi.$

Hence for either a star or a delta balanced connection the total power P is given by:

$P = \sqrt{3}\,V_LI_L \cos \phi$ **watts or** $P = 3I_p^2R_p$ **watts.**

Total volt-amperes, $S = \sqrt{3}\,V_LI_L$ volt-amperes.

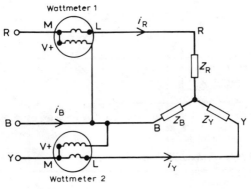

Figure 19.10

10   Power in three-phase loads may be measured by the following
methods:

(i)   *One-wattmeter method for a balanced load*
Wattmeter connections for both star and delta are shown in *Figure 19.9*. Total power = 3 × wattmeter reading.

(ii)   *Two-wattmeter method for balanced or unbalanced loads*
A connection diagram for this method is shown in *Figure 19.10* for a star-connected load. Similar connections are made for a delta-connected load.

Total power = sum of wattmeter readings = $P_1 + P_2$.

The power factor may be determined from:

$$\tan \phi = \sqrt{3} \frac{(P_1 - P_2)}{(P_1 + P_2)}$$

It is possible, depending on the load power factor, for one wattmeter to have to be 'reversed' to obtain a reading. In this case it is taken as a negative reading.

(iii)   *Three-wattmeter method for a three-phase, 4-wire system for balanced and unbalanced loads (see Figure 19.11)*

Total power = $P_1 + P_2 + P_3$.

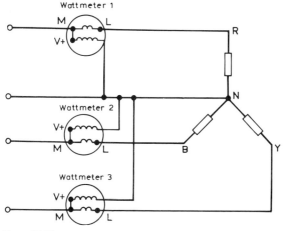

Figure 19.11

11 (i) Loads connected in delta dissipate three times more power than when connected in star to the same supply.

(ii) For the same power, the phase currents must be the same for both delta and star connections (since power $= 3I_p^2 R_p$), hence the line current in the delta-connected system is greater than the line current in the corresponding star-connected system. To achieve the same phase current in a star-connected system as in a delta-connected system, the line voltage in the star system is $\sqrt{3}$ times the line voltage in the delta system.

Thus for a given power transfer, a delta system is associated with larger line currents (and thus larger conductor cross-sectional area) and a star system is associated with a larger line voltage (and thus greater insulation).

12 **Advantages of three-phase systems over single-phase supplies** include:

(i) For a given amount of power transmitted through a system, the three-phase system requires conductors with a smaller cross-sectional area. This means a saving of copper (or aluminium) and thus the original installation costs are less.

(ii) Two voltages are available (see para. 7).

(iii) Three-phase motors are very robust, relatively cheap, generally smaller, have self-starting properties, provide a steadier output and require little maintenance compared with single-phase motors.

20 **d.c. transients**

Transients in series connected C-R circuits

1   When a d.c. voltage is applied to a capacitor C, and resistor R connected in series, there is a short period of time immediately after the voltage is connected, during which the current flowing in the circuit and the voltages across C and R are changing. These changing values are called transients.

Charging

2   (a) The circuit diagram for a series connected C–R circuit is shown in *Figure 20.1*. When switch S is closed, then by Kirchhoff's voltage law:

$$V = v_C + v_R \qquad (1)$$

(b) The battery voltage V is constant. The capacitor voltage c_C is given by q/C, where q is the charge on the capacitor. The voltage drop across R is given by iR, where i is the current flowing in the circuit. Hence, at all times:

$$V = \frac{q}{C} + iR \qquad (2)$$

At the instant of closing S (initial circuit condition), assuming there is no initial charge on the capacitor, q_0 is zero, hence v_{C_0} is zero. Thus from equation (1), $V = 0 + v_{R_0}$, i.e. $V_{R_0} = V$. This shows that the resistance to current is solely due to R, and the initial current flowing, $i_0 = I = V/R$.

(c) A short time later at time t_1 seconds after closing S, the capacitor is partly charged to, say, q_1 coulombs because current has been flowing. The voltage v_{C_1} is now q_1/C volts. If the current flowing is i_1 amperes, then the voltage drop across R has fallen to $i_1 R$ volts. Thus, equation (2) is now $V = (q_1/C) + i_1 R$.

(d) A short time later still, say at time t_2 seconds after closing the switch, the charge has increased to q_2 coulombs and v_C has increased q_2/C volts. Since $V = v_C + v_R$ and V is a

143

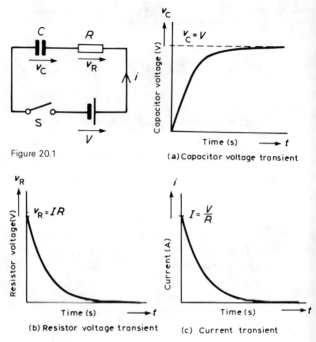

Figure 20.1

(a) Capacitor voltage transient

(b) Resistor voltage transient

(c) Current transient

Figure 20.2

constant, then v_R decreases to i_2R. Thus v_C is increasing and i and v_R are decreasing as time increases.

(e) Ultimately, a few seconds after closing S (final condition or steady state condition), the capacitor is fully charged to, say, Q coulombs, current no longer flows, i.e. $i=0$, and hence $v_R = iR = 0$. It follows from equation (1) that $v_C = V$.

(f) Curves showing the changes in v_C, v_R and i with time are shown in *Figure 20.2*. The curve showing the variation of v_C with time is called an **exponential growth curve** and the graph is called the 'capacitor voltage/time' characteristic. The curves showing the variation of v_R and i with time are called **exponential decay curves**, and the graphs are called 'resistor voltage/time' and 'current/time' characteristics respectively. (The name 'exponential' shows that the shape can be expressed mathematically by an exponential mathematical equation, see para. 5 below.)

The time constant

3   (a) If a constant d.c. voltage is applied to a series connected C–R circuit, a transient curve of capacitor voltage v_C is as shown in *Figure 20.2(a)*.

      (b) With reference to *Figure 20.3*, let the constant voltage supply be replaced by a variable voltage supply at time t_1 seconds. Let the voltage be varied so that the current flowing in the circuit is constant.

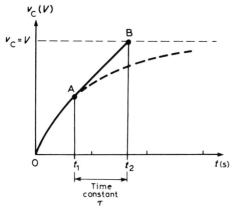

Figure 20.3

      (c) Since the current flowing is a constant, the curve will follow a tangent, AB, drawn to the curve at point A.

      (d) Let the capacitor voltage v_C reach its final value of V at time t_2 seconds.

      (e) The time corresponding to $(t_2 - t_1)$ seconds is called the time constant of the circuit, denoted by the Greek letter 'tau', τ. The value of the time constant is CR seconds, i.e. for a series connected C–R circuit, **time constant, $\tau = CR$ seconds**.

        Since the variable voltage mentioned in para. 3(b) above can be applied at any instant during the transient change, it may be applied at $t=0$, i.e. at the instant of connecting the circuit to the supply. If this is done, then the time constant of the circuit may be defined as:

'*the time taken for a transient to reach its final state if the initial rate of change is maintained*'.

4   There are two main methods of drawing transient curves graphically:

(a) **The tangent method**.

(b) **The initial slope and three point method** which is based on the following properties of a transient exponential curve:

(i) For a growth curve, the value of a transient at a time equal to one time constant is 0.632 of its steady state value (usually taken as 63% of the steady state value); at a time equal to two and a half time constants is 0.918 of its steady state value (usually taken as 92% of its steady state value) and at a time equal to five time constants is equal to its steady state value.

(ii) For a decay curve, the value of a transient at a time equal to one time constant is 0.368 of its initial value (usually taken as 37% of its initial value), at a time equal to two and a half time constants is 0.082 of its initial value (usually taken as 8% of its initial value) and at a time equal to five time constants is equal to zero.

5   The transient curves shown in *Figure 20.2* have mathematical equations, obtained by solving the differential equations representing the circuit. The equations of the curves are:

growth of capacitor voltage, $v_C = V(1 - e^{-t/CR}) = V(1 - e^{-t/\tau})$,

decay of resistor voltage, $v_R = Ve^{(-t/CR)} = Ve^{(-t/\tau)}$ and

decay of current flowing, $i = Ie^{(-t/CR)} = Ie^{(-t/\tau)}$.

For example, let a 15 μF uncharged capacitor be connected in series with a 47 kΩ resistor across a 120 V, d.c. supply. To draw the capacitor voltage/time characteristic of the circuit using the tangential graphical method, the time constant of the circuit and the steady state value need first to be determined.

Time constant $= CR = 15$ μF $\times 47$ k$\Omega = 15 \times 10^{-6} \times 47 \times 10^3$
                                                        $= 0.705$ s.

Steady state value of v_C is $v_C = V$, i.e. $v_C = 120$ V.

With reference to *Figure 20.4*, the scale of the horizontal axis is drawn so that it spans at least five time constants, i.e. 5×0.705 or about $3\frac{1}{2}$ seconds. The scale of the vertical axis spans the change in

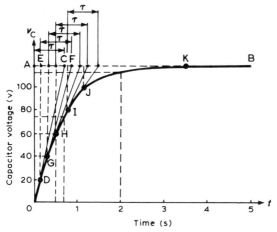

Figure 20.4

the capacitor voltage, that is, from 0 to 120 V. A broken line AB is drawn corresponding to the final value of v_C.

Point C is measured along AB so that AC is equal to 1τ, i.e. AC=0.705 s. Straight line OC is drawn. Assuming that about five intermediate points are needed to draw the curve accurately, a point D is selected on OC corresponding to a v_C value of about 20 V. DE is drawn vertically. EF is made to correspond to 1τ, i.e. EF=0.705 s. A straight line is drawn joining DF. This procedure of

  (a) drawing a vertical line through point selected,
  (b) at the steady-state value, drawing a horizontal line corresponding to 1τ, and
  (c) joining the first and last points,

is repeated for v_C values of 40, 60, 80 and 100 V, giving points G, H, I and J.

The capacitor voltage effectively reaches its steady-state value of 120 V after a time equal to five time constants, shown as point K. Drawing a smooth curve through points O, D, G, H, I, J and K gives the exponential growth curve of capacitor voltage.

From the graph, the value of capacitor voltage at a time equal to the time constant is about 75 V. It is a characteristic of all exponential growth curves, that after a time equal to one time constant, the value of the transient is 0.632 of its steady-state value. In this problem, $0.632 \times 120 = 75.84$ V. Also from the graph, when t

is two seconds, v_C is about 115 V. (This value may be checked using the equation $v_C = V(1 - e^{-t/\tau})$, where $V = 120$ V, $\tau = 0.705$ s and $t = 2$ s. This calculation gives $v_C = 112.97$ V.)

The time for v_C to rise to one half of its final value, i.e. 60 V, can be determined from the graph and is about 0.5 s. This value may be checked using $v_C = V(1 - e^{-t/\tau})$ where $V = 120$ V, $v_C = 60$ V and $\tau = 0.705$ s, giving $t = 0.489$ s.

As another example, let a 4 μF capacitor be charged to 24 V and then discharged through a 220 kΩ resistor. To draw the capacitor voltage/time and the current/time characteristics using the 'initial slope and three point' method, the time constant of the circuit and the steady state value are firstly needed.

Time constant, $\tau = CR = 4 \times 10^{-6} \times 220 \times 10^3 = 0.88$ s.

Initially, capacitor voltage $v_C = v_R = 24$ V.

$$i = \frac{V}{R} = \frac{24}{220 \times 10^3} = 0.109 \text{ mA.}$$

Finally, $v_C = v_R = i = 0$.

The exponential decay of capacitor voltage is from 24 V to 0 V in a time equal to five time constants, i.e. $5 \times 0.88 = 4.4$ s. With reference to *Figure 20.5*, to construct the decay curve:

  (i)   the horizontal scale is made so that it spans at least five time constants, i.e. 4.4 s,

  (ii)   the vertical scale is made to span the change in capacitor voltage, i.e. 0 to 24 V,

  (iii) point A corresponds to the initial capacitor voltage, i.e. 24 V,

  (iv) OB is made equal to one time constant and line AB is drawn. This gives the initial slope of the transient,

  (v)   the value of the transient after a time equal to one time constant is 0.368 of the initial value, i.e.

$$0.368 \times 24 = 8.83 \text{ V.}$$

A vertical line is drawn through B and distance BC is made equal to 8.83 V,

  (vi) the value of the transient after a time equal to two and a half time constants is 0.082 of the initial value, i.e.

$$0.082 \times 24 = 1.97 \text{ V,}$$

shown as point D in *Figure 20.5*,

  (vii) the transient effectively dies away to zero after a time equal to five time constants, i.e. 4.4 s, giving point E.

The smooth curve drawn through points A, C, D and E represents

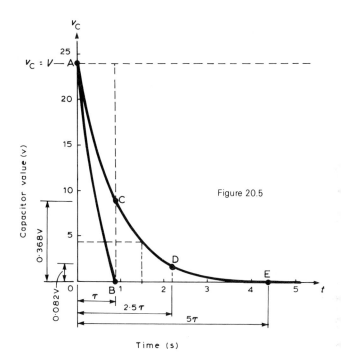

Figure 20.5

Time (s)

he decay transient. At $1\frac{1}{2}$ s after decay has started, $v_C \approx 4.4$ V. This may be checked using $v_C = Ve^{-t/\tau}$, where $V = 24$, $t = 1\frac{1}{2}$ and $\tau = 0.88$, giving $v_C = 4.36$ V.

The voltage drop across the resistor is equal to the capacitor voltage when a capacitor is discharging through a resistor, thus the resistor voltage/time characteristic is identical to that shown in *Figure 20.5*. Since $v_R = v_C$, then at $1\frac{1}{2}$ seconds after decay has started,

$v_R \approx 4.4$ V (see above).

The current/time characteristic is constructed in the same way as the capacitor·voltage/time characteristic, and is as shown in *Figure 20.6*. The values are:

point A: initial value of current = 0.109 mA
point C: at 1τ, $i = 0.368 \times 0.109 = 0.040$ mA
point D: at 2.5τ, $i = 0.082 \times 0.109 = 0.009$ mA
point E: at 5τ, $i = 0$.

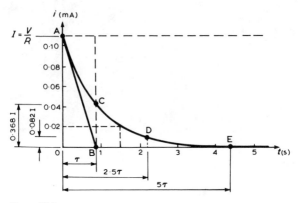

Figure 20.6

Hence current transient is as shown. At a time of $1\frac{1}{2}$ seconds, the value of current, from the characteristic is 0.02 mA. This may be checked using $i = Ie^{(-t/\tau)}$ where $I = 0.109$, $t = 1\frac{1}{2}$ and $\tau = 0.88$, giving $i = 0.0198$ mA or 19.8 μA.

Discharging

6   When a capacitor is charged (i.e. with the switch in position A in *Figure 20.7*), and the switch is then moved to position B, the electrons stored in the capacitor keep the current flowing for a short time. Initially, at the instant of moving from A to B, the current flow is such that the capacitor voltage v_C is balanced by an equal and opposite voltage $v_R = iR$.

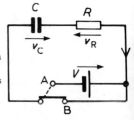

Figure 20.7

Since initially $v_C = v_R = V$, then $i = I = V/R$.

During the transient decay, by applying Kirchhoff's voltage law to *Figure 20.8* $v_C = v_R$. Finally the transients decay exponentially to zero, i.e. $v_C = v_R = 0$. The transient curves representing the voltages and current are as shown in *Figure 20.8*.

7   The equations representing the transient curves during the

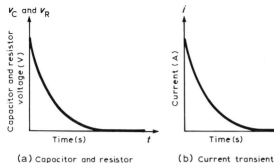

Figure 20.8

discharge period of a series connected C–R circuit are:

decay of voltage, $v_C = v_R = V e^{(-t/CR)} = V e^{(-t/\tau)}$

decay of current, $i = I e^{(-t/CR)} = I e^{(-t/\tau)}$

Transients in series connected L-R circuits

8   When a d.c. voltage is connected to a circuit having inductance L connected in series with resistance R, there is a short period of time immediately after the voltage is connected, during which the current flowing in the circuit and the voltages across L and R are changing. These changing values are called transients.

9   **Current growth**
   (a) The circuit diagram for a series connected L–R circuit is shown in *Figure 20.9*. When switch S is closed, then by Kirchhoff's voltage law:
   $$V = v_L + v_R \qquad (3)$$

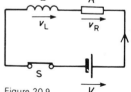

Figure 20.9

   (b) The battery voltage V is constant. The voltage of the inductance is the induced voltage, i.e.

   $$v_L = L \times \frac{\text{change of current}}{\text{change of time}} \qquad (4)$$

shown as $L(di/dt)$. The voltage drop across R, v_R is given by iR. Hence, at all times: $V = L(di/dt) + iR$.

(c) At the instant of closing the switch, the rate of change of current is such that it induces an e.m.f. in the inductance which is equal and opposite to V, hence $V = v_L + 0$, i.e. $v_L = V$. From equation (3), because $v_L = V$, then $v_R = 0$ and $i = 0$.

(d) A short time later at time t_1 seconds after closing S, current i_1 is flowing, since there is a rate of change of current initially, resulting in a voltage drop of $i_1 R$ across the resistor. Since V (constant) $= v_L + v_R$ the induced e.m.f. is reduced and equation (4) becomes:

$$V = L \frac{di_1}{dt_1} + i_1 R.$$

(e) A short time later still, say at time t_2 seconds after closing the switch, the current flowing is i_2, and the voltage drop across the resistor increases to $i_2 R$. Since v_R increases, v_L decreases.

(f) Ultimately, some time after closing S, the current flow is entirely limited by R, the rate of change of current is zero and hence v_L is zero. Thus $V = iR$. Under these conditions, steady state current flows, usually signified by I. Thus, $I = V/R$, $v_R = IR$ and $v_L = 0$, at steady state conditions.

(g) Curves showing the changes in v_L, v_R and i with time are shown in *Figure 20.10* and indicate that v_L is a maximum value initially (i.e. equal to V), decaying exponentially to zero, whereas v_R and i grow from zero to their steady state values of V and $I = V/R$ respectively.

The time constant
10   With reference to para. 3, the time constant of a series connected L–R circuit is defined in the same way as the time constant for a series connected C–R circuit. Its value is given by:

time constant, $\tau = L/R$ **seconds**.

11   Transient curves representing the induced voltage/time, resistor voltage/time and current/time characteristics may be drawn graphically, as outlined in para. 4.

12   Each of the transient curves shown in *Figure 20.10* have mathematical equations, and these are:

decay of induced voltage, $v_L = Ve^{(-Rt/L)} = Ve^{(-t/\tau)}$,

152

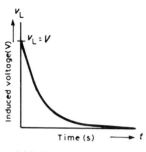

(a) Induced voltage transient

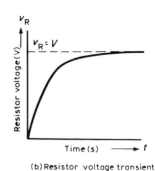

(b) Resistor voltage transient

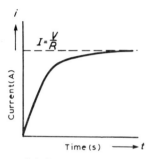

(c) Current transient

Figure 20.10

growth of resistor voltage, $v_R = V(1 - e^{(-Rt/L)}) = V(1 - e^{-t/\tau})$,

growth of current flow, $i = I(1 - e^{(-Rt/L)}) = I(1 - e^{-t/\tau})$.

Current decay

13   When a series connected L-R circuit is connected to a d.c. supply as shown with S in position A of *Figure 20.11*, a current $I = V/R$ flows after a short time, creating a magnetic field $(\Phi \propto I)$ associated with the inductor. When S is moved to position B, the current value decreases, causing a decrease in the strength of the magnetic field. Flux linkages occur, generating a voltage v_L, equal to $L(di/dt)$. Thus $v_L = v_R$.

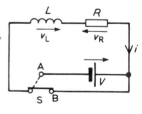

Figure 20.11

153

The current decays exponentially to zero and since v_R is proportional to the current flowing, v_R decays exponentially to zero. Since $v_L = v_R$, v_L also decays exponentially to zero. The curves representing these transients are similar to those shown in *Figure 20.8*.

14   The equations representing the decay transient curves are:

decay of voltages, $v_L = v_R = Ve^{(-Rt/L)} = Ve^{(-t/\tau)}$

decay of current, $i = Ie^{(-Rt/L)} = Ie^{(-t/\tau)}$

The effects of time constant on a rectangular wave

15   By varying the value of either C or R in a series connected C-R circuit, the time constant $(\tau = CR)$, of a circuit can be varied. If a rectangular waveform varying from $+E$ to $-E$ is applied to a C-R circuit, as shown in *Figure 20.12*, output waveforms of the capacitor

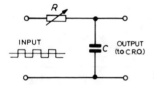

INPUT

OUTPUT (to C.R.O.)

Figure 20.12

voltage have various shapes, depending on the value of R. When R is small, $\tau = CR$ is small and an output waveform such as that shown in *Figure 20.13(a)* is obtained.

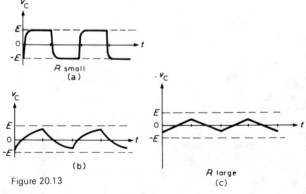

Figure 20.13

As the value of R is increased, the waveform changes to that shown in *Figure 20.13(b)*. When R is large, the waveform is as shown in *Figure 20.13(c)*, the circuit then being described as an integrator circuit.

16   If a rectangular waveform varying from $+E$ to $-E$ is applied to a series connected C-R circuit and the waveform of the voltage drop across the resistor is observed, as shown in *Figure 20.14*, the output waveform alters as R is varied due to the time constant, $(\tau = CR)$, altering. When R is small, the waveform is as shown in *Figure 20.15(a)*, the voltage being generated across R by the capacitor discharging fairly quickly.

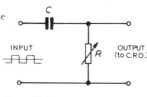

Figure 20.14

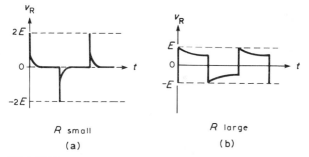

Figure 20.15

Since the change in capacitor voltage is from $+E$ to $-E$, the change in discharge current is $2E/R$, resulting in a change in voltage across the resistor of $2E$. This circuit is called a differentiator circuit. When R is large, the waveform is as shown in *Figure 20.15(b)*.

21 Single phase transformers

1   **Mutual inductance, M**, is the property whereby an e.m.f. is induced in a circuit by a change of flux due to current changing in an adjacent circuit.

2   (i) A **transformer** is a device which uses the phenomenon of mutual induction to change the values of alternating voltages and currents. In fact, one of the main advantages of a.c. transmission and distribution is the ease with which an alternating voltage can be increased or decreased by transformers.

(ii) **Losses** in transformers are generally low and thus efficiency is high. Being static they have a long life and are very reliable.

(iii) Transformers range in size from the miniature units used in electronic applications to the large power transformers used in power stations. The principle of operation is the same for each.

3   A transformer is represented in *Figure 21.1(a)* as consisting of two electrical circuits linked by a common ferromagnetic core. One coil is termed the **primary winding** which is connected to the

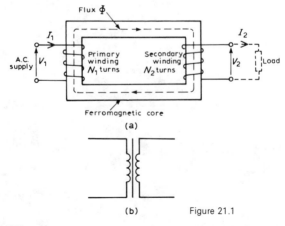

(a)

(b)   Figure 21.1

supply of electricity, and the other the **secondary winding**, which may be connected to a load. A circuit diagram symbol for a transformer is shown in *Figure 21.1(b)*.

Transformer principle of operation

4   (i) When the secondary is an open-circuit and an alternating voltage V_1 is applied to the primary winding, a small current — called the no-load current I_o — flows, which sets up a magnetic flux in the core. This alternating flux links with both primary and secondary coils and induces in them emfs of E_1 and E_2 volts respectively by mutual induction.

(ii) The induced e.m.f. E in a coil of N turns is given by $E = N(\Delta\Phi)/t$ volts, where $(\Delta\Phi/t)$ is the rate of change of flux. In an ideal transformer, the rate of change of flux is the same for both primary and secondary and thus $E_1/N_1 = E_2/N_2$, i.e. **the induced e.m.f. per turn is constant**.

Assuming no losses, $E_1 = V_1$ and $E_2 = V_2$.

$$\text{Hence } \frac{V_1}{N_1} = \frac{V_2}{N_2} \text{ or } \frac{V_1}{V_2} = \frac{N_1}{N_2} \tag{1}$$

(iii) V_1/V_2 is called the **voltage ratio** and N_1/N_2 the **turns ratio**, or the '**transformation ratio**' of the transformer.

If N_2 is less than N_1 then V_2 is less than V_1 and the device is termed a **step-down transformer**.

If N_2 is greater than N_1, then V_2 is greater than V_1 and the device is termed a **step-up transformer**.

(iv) When a load is connected across the secondary winding, a current I_2 flows. In an ideal transformer losses are neglected and a transformer is considered to be 100% efficient. Hence

input power = output power
i.e. $V_1 I_1 \cos \phi_1 = V_2 I_2 \cos \phi_2$

However the primary and secondary power factors (i.e. $\cos \phi_1$ and $\cos \phi_2$) are nearly equal at full load. Hence, $V_1 I_1 = V_2 I_2$, i.e. in an ideal transformer, the primary and secondary ampere-turns are equal.

$$\text{Thus } \frac{V_1}{V_2} = \frac{I_2}{I_1} \tag{2}$$

Combining equations (1) and (2) gives:

$$\boxed{\frac{V_1}{V_2} = \frac{N_1}{N_2} = \frac{I_2}{I_1}} \tag{3}$$

5   The **rating** of a transformer is stated in terms of the volt-amperes that it can transform without overheating. With reference to *Figure 21.1(a)*, the transformer rating is either V_1I_1 or V_2I_2, where I_2 is the full-load secondary current.

6   There are broadly two sources of losses in transformers on load-copper losses and iron losses.

(a) **Copper losses** are variable and result in a heating of the conductors, due to the fact that they possess resistance. If R_1 and R_2 are the primary and secondary winding resistances then the total copper loss is $I_1^2 R_1 + I_2^2 R_2$.

(b) **Iron losses** are constant for a given value of frequency and flux density and are of two types — hysteresis loss and eddy current loss.

(i) **Hysteresis loss** is the heating of the core as a result of the internal molecular structure reversals which occur as the magnetic flux alternates. The loss is proportional to the area of the hysteresis loop and thus low loss nickel iron alloys are used for the core since their hysteresis loops have small areas.

(ii) **Eddy current loss** is the heating of the core due to e.m.f.'s being induced not only in the transformer windings but also in the core. These induced e.m.f.'s set up circulating currents, called eddy currents. Owing to the low resistance of the core eddy currents can be quite considerable and can cause a large power loss and excessive heating of the core. Eddy current losses can be reduced by increasing the resistivity of the core material or, more usually, by laminating the core (i.e. splitting it into layers or leaves) when very thin layers of insulating material can be inserted between each pair of laminations. This increases the resistance of the eddy current path, and reduces the value of the eddy current.

7   Transformer efficiency,

$$\eta = \frac{\text{output power}}{\text{input power}} = \frac{\text{input power-losses}}{\text{input power}}$$

$$= 1 - \frac{\textbf{losses}}{\textbf{input power}}$$

and is usually expressed as a percentage. It is not uncommon for power transformers to have efficiencies of between 95% and 98%. The output power $= V_2I_2 \cos \phi_2$, the losses $=$ copper loss $+$ iron losses, and the input power $=$ output power $+$ losses.

Transformer no-load phasor diagram

8 (i) The core flux is common to both primary and secondary windings in a transformer and is thus taken as the reference phasor in a phasor diagram. On no-load the primary winding takes a small no-load current I_o and since, with losses neglected, the primary winding is a pure inductor, this current lags the applied voltage V_1 by 90°. In the phasor diagram assuming no losses, shown in *Figure 21.2(a)*, current I_o produces the flux and is drawn in phase with the flux. The primary induced e.m.f. E_1 is in phase opposition to V_1 (by Lenz's law) and is shown 180° out of phase with V_1 and equal in magnitude. The secondary induced e.m.f. is shown for a 2:1 turns ratio transformer.

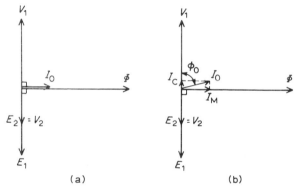

(a)

(b)

Figure 21.2

(ii) A no-load phasor diagram for a practical transformer is shown in *Figure 21.2(b)*. If current flows then losses will occur. When losses are considered then the no-load current I_o is the phasor sum of two components — (i) I_M, **the magnetising component**, in phase with the flux, and (ii) I_c, the **core loss component** (supplying the hysteresis and eddy current losses). From *Figure 21.2(b)*:

No-load current, $I_o = \sqrt{(I_M^2 + I_c^2)}$.

Power factor on no-load $= \cos \phi_0 = \dfrac{I_c}{I_0}$

The total core losses (i.e. iron losses) $= V_1 I_0 \cos \phi_0$.

Transformer construction

9 (i) There are broadly two types of transformer construction — the **core type** and the **shell type**, as shown in *Figure 21.3*. The low and high voltage windings are wound as shown to reduce leakage flux.

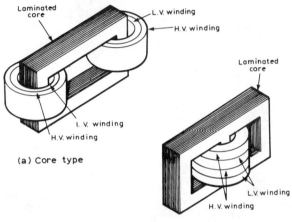

Figure 21.3

(a) Core type

(b) Shell type

(ii) For **power transformers**, rated possibly at several MVA and operating at a frequency of 50 Hz in Great Britain, the core material used is usually laminated silicon steel or stalloy, the laminations reducing eddy currents and the silicon steel keeping hysteresis loss to a minimum.

(iii) For **audio-frequency (a.f.) transformers**, rated from a few mVA to no more than 20 VA, and operating at frequencies up to about 15 kHz, the small core is also made of laminated silicon steel.

(iv) **Radio-frequency (r.f.) transformers**, operating in the MHz frequency region have either an air core, a ferrite core or a dust core. Ferrite is a ceramic material having magnetic properties similar to silicon steel, but having a high resistivity. Dust cores consist of fine particles of carbonyl iron or permalloy (i.e. nickel and iron), each particle of which is insulated from its neighbour.

(v) Transformer **windings** are usually of enamel-insulated copper or aluminium.

(vi) **Cooling** is achieved by air in small transformers and oil in large transformers.

10   The **maximum transfer theorem** states:

'The power transferred from a supply source to a load is at its maximum when the resistance of the load is equal to the internal resistance of the source'.

In *Figure 21.4*, when $r = R$, the power transferred from the source to the load is a maximum.

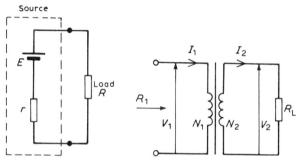

Figure 21.4   Figure 21.5

11   A method of achieving maximum power transfer between a source and a load is to adjust the value of the load resistance to 'match' the source internal resistance. A transformer may be used as a **resistance matching device** by connecting it between the load and the source. The reason why a transformer can be used for this is shown below. With reference to *Figure 21.5*:

$$R_L = \frac{V_2}{I_2} \text{ and } R_1 = \frac{V_1}{I_1}$$

For an ideal transformer, $V_1 = (N_1/N_2) V_2$ and $I_1 = (N_2/N_1) I_2$ from equation (3)). Thus the equivalent input resistance R_1 of the transformer is given by:

$$R_1 = \frac{V_1}{I_1} = \frac{\left(\dfrac{N_1}{N_2}\right)V_2}{\left(\dfrac{N_2}{N_1}\right)I_2} = \left(\frac{N_1}{N_2}\right)^2 \frac{V_2}{I_2} = \left(\frac{N_1}{N_2}\right)^2 R_L$$

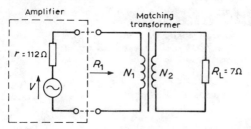

Figure 21.6

Hence by varying the value of the turns ratio, the equivalent input resistance a transformer can be 'matched' to the internal resistance of a load to achieve power transfer.

For example, the output stage of an amplifier has an output resistance of $112\,\Omega$. The optimum turns ratio of a transformer which would match a load resistance of $7\,\Omega$ to the output resistance of the amplifier is obtained as follows:

The equivalent input resistance R_1 of the transformer needs to be $112\,\Omega$ for maximum power transfer. The circuit diagram is shown in *Figure 21.6*.

$$R_1 = \left(\frac{N_1}{N_2}\right)^2 R_L$$

Hence $\left(\dfrac{N_1}{N_2}\right)^2 = \dfrac{R_1}{R_L} = \dfrac{112}{7} = 16$

i.e. $\dfrac{N_1}{N_2} = \sqrt{16} = 4$

Hence the optimum turns ratio is 4:1.

22 d.c. machines

1   When the input to an electrical machine is electrical energy (seen as applying a voltage to the electrical terminals of the machine), and the output is mechanical energy (seen as a rotating shaft), the machine is called an **electric motor**. Thus an electric motor converts electrical energy into mechanical energy.

2   When the input to an electrical machine is mechanical energy (seen as, say, a diesel motor, coupled to the machine by a shaft), and the output is electrical energy (seen as a voltage appearing at the electrical terminals of the machine), the machine is called a **generator**. Thus, a generator converts mechanical energy to electrical energy.

3   The efficiency of an electrical machine is the ratio of the output power to the input power and is usually expressed as a percentage. The Greek letter 'eta', 'η' is used to signify efficiency and since the units are power/power, then efficiency has no units. Thus:

$$\textbf{efficiency}, \eta = \frac{\textbf{output power}}{\textbf{input power}} \times \textbf{100}\%.$$

4   **The action of a commutator**. In an electric motor, conductors rotate in a uniform magnetic field. A single-loop conductor mounted between permanent magnets is shown in *Figure 22.1*. A voltage is applied at points A and B in *Figure 22.1(a)*.

A force, F, acts on the loop due to the interaction of the magnetic field of the permanent magnets and the magnetic field created by the current flowing in the loop. This force is proportional to the flux density, B, the current flowing, I, and the effective length of the conductor, l, i.e. $F = BIl$. The force is made up of two parts, one acting vertically downwards due to the current flowing from C to D and the other acting vertically upwards due to the current flowing from E to F (from Fleming's left hand rule). If the loop is free to rotate, then when it has rotated through 180°, the conductors are as shown in *Figure 22.1(b)*. For rotation to continue in the same direction, it is necessary for the current flow to be as shown in *Figure 22.1(b)*, i.e.

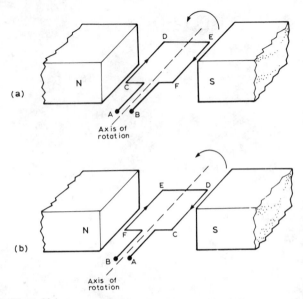

Figure 22.1

from D to C and from F to E. This apparent reversal in the direction of current flow is achieved by a process called **commutation**. With reference to *Figure 22.2(a)*, when a direct voltage is applied at A and B, then as the single-loop conductor rotates, current flow will always be away from the commutator for the part of the conductor adjacent to the N-pole and towards the

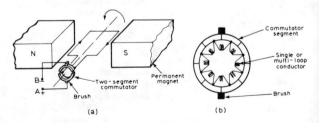

Figure 22.2

commutator for the part of the conductor adjacent to the S-pole. Thus the forces act to give continuous rotation in an anti-clockwise direction. The arrangement shown in *Figure 22.2* is called a 'two-segment' commutator and the voltage is applied to the rotating segments by stationary **brushes** (usually carbon blocks), which slide on the commutator material (usually copper), when rotation takes place.

In practice, there are many conductors on the rotating part of a d.c. machine and these are attached to many commutator segments. A schematic diagram of a multi-segment commutator is shown in *Figure 22.2(b)*.

5   **D.C. machine construction**. The basic parts of any d.c. machine are shown in *Figure 22.3(b)*, and comprise:

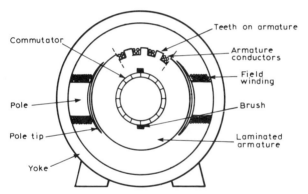

Figure 22.3

(a) a stationary part called the **stator** having,
   (i) a steel ring called the **yoke**, to which are attached;
   (ii) the magnetic **poles**, around which are the;
   (iii) **field windings**, i.e. many turns of a conductor wound round the pole core.
   Current passing through this conductor creates an electromagnet (rather than the permanent magnets shown in *Figures 22.1* and *22.2*).
(b) A rotating part called the **armature** mounted in bearings housed in the stator and having,
   (iv) a laminated cylinder of iron or steel called the **core**, on which teeth are cut to house the

165

(v) **armature winding**, i.e. a single or multi-loop conductor system and

(vi) the **commutator** (see para. 4).

6   The average e.m.f. induced in a single conductor on the armature of a d.c. machine is given by:

$$\frac{\text{flux cut/rev}}{\text{time of 1 rev}} = \frac{2p\Phi}{1/n}$$

where p is the number of pairs of poles, Φ is the flux in webers entering or leaving a pole and n is the speed of rotation in rev/s. Thus the average e.m.f. per conductor is $2p\Phi n$ volts. If there are Z conductors connected in series, the average e.m.f. generated is $2p\Phi n Z$ volts. For a given machine, the number of pairs of poles p and the number of conductors connected in series Z are constant, hence the generated e.m.f. is proportional to Φn. But $2\pi n$ is the angular velocity, ω in rad/s, hence the generated e.m.f. E is proportional to Φ and to ω, i.e. generated e.m.f.,

$$E \propto \Phi\omega \tag{1}$$

7   The power on the shaft of a d.c. machine is the product of the torque and the angular velocity, i.e.

shaft power $= T\omega$ **watts**

where T is the torque in Nm and ω is the angular velocity in rad/s. The power developed by the armature current is EI_a watts, where E is the generated e.m.f. in volts and I_a is the armature current in amperes. If losses are neglected then $T\omega = EI_a$. But from para. 6, $E \propto \Phi\omega$.

Hence $T\omega \propto \Phi\omega I_a$, i.e. $T \propto \Phi I_a$ $\tag{2}$

8   When the field winding of a d.c. machine is connected in parallel with the armature, as shown in *Figure 22.4(a)*, the machine is said to be **shunt wound**.

If the field winding is connected in series with the armature, as shown in *Figure 22.4(b)*, then the machine is said to be **series wound**.

9   Depending on whether the electrical machine is series wound or shunt wound, it behaves differently when a load is applied. The behaviour of a d.c. machine under various conditions is shown by means of graphs, called characteristic curves or just characteristics. The characteristics shown in the paras below are theoretical, since they neglect the effects of such things as armature reaction and demagnetising ampere-turns.

166

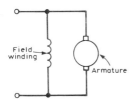

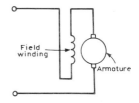

(a) Shunt-wound machine   (b) Series-wound machine

Figure 22.4

Shunt-wound motor characteristics

10   The two principal characteristics are the torque/armature
current and speed/armature current relationships. From these, the
torque/speed relationship can be derived.

(i) The theoretical torque/armature current characteristic can be
derived from the expression $T \propto \Phi I_a$ (see para. 7). For a shunt-
wound motor, the field winding is connected in parallel with the
armature circuit and thus the applied voltage gives a constant field
current, i.e. a shunt-wound motor is a constant flux machine. Since
Φ is constant, it follows that $T \propto I_a$, and the characteristic is as
shown in *Figure 22.5(a)*.

(ii) The armature circuit of a d.c. motor has resistance due to the
armature winding and brushes, R_a ohms, and when armature
current I_a is flowing through it, there is a voltage drop of $I_a R_a$ volts.
In *Figure 22.5(b)* the armature resistance is shown as a separate
resistor in the armature circuit to help understanding. Also, even
though the machine is a motor, because conductors are rotating in
a magnetic field, a voltage, $E \propto \Phi \omega$, is generated by the armature
conductors. By applying Kirchhoff's voltage law to the armature
circuit ABCD in *Figure 22.5(b)*, the voltage equation is $V = E + I_a R_a$,
i.e. $E = V - I_a R_a$. But from para. 6, $E \propto \Phi n$, hence $n \propto E/\Phi$, i.e.
speed of rotation,

$$n \propto \frac{E}{\Phi} \propto \frac{V - I_a R_a}{\Phi} \tag{3}$$

For a shunt motor, V, Φ and R_a are constants, hence as armature
current I_a increases, $I_a R_a$ increases and $V - I_a R_a$ decreases, and the
speed is proportional to a quantity which is decreasing and is as
shown in *Figure 22.5(c)*. As the load on the shaft of the motor
increases, I_a increases and the speed drops slightly. In practice, the

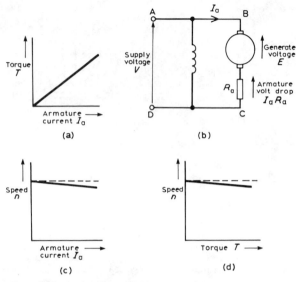

Figure 22.5

speed falls by about 10% between no-load and full-load on many d.c. shunt-wound motors.

(iii) Since torque is proportional to armature current (see (i) above), the theoretical speed/torque characteristic is as shown in *Figure 22.5(d)*.

Series-wound motor characteristics

11   In a series motor, the armature current flows in the field winding and is equal to the supply current I (see *Figure 22.4(b)*).
(i) **The torque/current characteristic** It is shown in para. 7 that torque $T \propto \Phi I_a$. Since the armature and field currents are the same current, I, in a series machine, then $T \propto \Phi I$ over a limited range, before magnetic saturation of the magnetic circuit of the motor is reached (i.e. the linear portion of the B-H curve for the yoke, poles, air gap, brushes and armature in series). Thus $\Phi \propto I$ and $T \propto I^2$. After magnetic saturation, Φ almost becomes a constant and $T \propto I$. Thus the theoretical torque/current characteristic is as shown in *Figure 22.6(a)*.

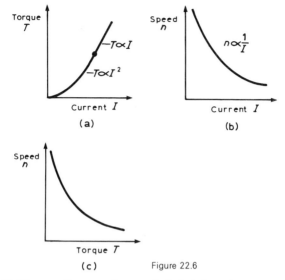

Figure 22.6

(ii) **The speed/current characteristic** It is shown in para. 10(ii) that $n \propto (V - I_a R_a)/\Phi$. In a series motor, $I_a = I$ and below the magnetic saturation level, $\Phi \propto I$. Thus $n \propto (V - IR)/I$ where R is the combined resistance of the series field and armature circuit. Since IR is small compared with V, then an approximate relationship for the speed is $n \propto 1/I$. Hence the theoretical speed/current characteristic is as shown in *Figure 22.6(b)*. The high speed at small values of current indicate that this type of motor must not be run on very light loads and invariably, such motors are permanently coupled to their loads.

(iii) The theoretical speed/torque characteristic may be derived from (i) and (ii) above by obtaining the torque and speed for various values of current and plotting the co-ordinates on the speed/torque characteristic. A typical speed/torque characteristic is shown in *Figure 22.6(c)*.

Shunt-wound generator characteristics

12 The two principal generator characteristics are the generated voltage/field current characteristic, called the open-circuit characteristic and the terminal voltage/load current characteristic, called the load characteristic.

(i) **The theoretical open-circuit characteristic** The generated e.m.f., E, is proportional to $\Phi\omega$ (see para. 6), hence at constant speed, since $\omega = 2\pi n$, $E \propto \Phi$. Also the flux Φ is proportional to field current I_f until magnetic saturation of the iron circuit of the generator occurs. Hence the open circuit characteristic is as shown in *Figure 22.7(a)*.

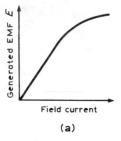

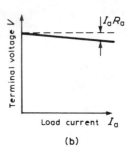

Figure 22.7

(ii) **The theoretical load characteristic** As the load current on a generator having constant field current and running at constant speed increases, the value of armature current increases, hence the armature volt drop, $I_a R_a$ increases. The generated voltage E is larger than the terminal voltage V and the voltage equation for the armature circuit is $V = E - I_a R_a$. Since E is constant, V decreases with increasing load. The load characteristic is as shown in *Figure 22.7(b)*. In practice, the fall in voltage is about 10% between no-load and full-load for many d.c. shunt-wound generators.

Series-wound generator characteristic

13 The load characteristic is the terminal voltage/current characteristic. The generated e.m.f., E, is proportional to $\Phi\omega$ and at constant speed ω $(= 2\pi n)$ is a constant. Thus E is proportional to Φ. For values of current below magnetic saturation of the yoke, poles, air gaps and armature core, the flux Φ is proportional to the current, hence $E \propto I$.

For values of current above those required for magnetic saturation, the generated e.m.f. is approximately constant. The values of field resistance and armature resistance in a series wound machine are small, hence the terminal voltage V is very nearly equal to E. Thus the theoretical load characteristic is similar in shape to the characteristic shown in *Figure 22.7(a)*.

In a series-wound generator, the field winding is in series with the armature and it is not possible to have a value of field current when the terminals are open circuited, thus it is not possible to obtain an open-circuit characteristic.

The d.c. motor starter

14   If a d.c. motor whose armature is stationary is switched directly to its supply voltage, it is likely that the fuses protecting the motor will burn out. This is because the armature resistance is small, frequently being less than $1\,\Omega$. Thus, additional resistance must be added to the armature circuit at the instant of closing the switch to start the motor.

As the speed of the motor increases, the armature conductors are cutting flux and a generated voltage, acting in opposition to the applied voltage, is produced, which limits the flow of armature current. Thus the value of the additional armature resistance can then be reduced. When at normal running speed, the generated e.m.f. is such that no additional resistance is required in the armature circuit. To achieve this varying resistance in the armature circuit on starting, a d.c. motor starter is used, as shown in *Figure 22.8*.

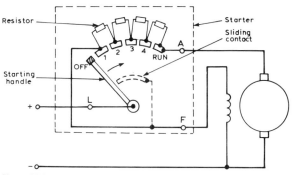

Figure 22.8

The starting handle is moved slowly in a clockwise direction to start the motor. For a shunt-wound motor, the field winding is connected to stud 1 or to L, via a sliding contact on the starting handle, to give maximum field current, hence maximum flux, hence maximum torque on starting, since $T \propto \Phi I_a$. A similar arrangement without the field connection is used for series motors.

Speed control of d.c. motors

15 (i) **Shunt-wound motors**

The speed of a shunt-wound d.c. motor, n, is proportional to $(V-I_aR_a)\Phi$ (see para. 10). The speed is varied either by varying the value of flux, Φ, or by varying the value of R_a. The former is achieved by using a variable resistor in series with the field winding, as shown in *Figure 22.9(a)*, and such a resistor is called the shunt field regulator. As the value of resistance of the shunt

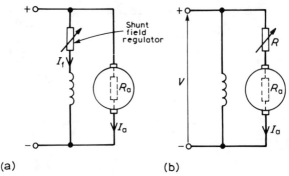

(a)   (b)

Figure 22.9

field regulator is increased, the value of the field current, I_f, is decreased. This results in a decrease in the value of flux Φ, and hence an increase in the speed, since $n \propto 1/\Phi$. Thus only speeds above that given without a shunt field regulator can be obtained by this method. Speeds below those given by $(V-I_aR_a)/\Phi$ are obtained by increasing the resistance in the armature circuit, as shown in *Figure 22.9(b)*, where

$$n \propto \frac{V-I_a(R_a+R)}{\Phi}$$

Since resistor R is in series with the armature, it carries the full armature current and results in a large power loss in large motors where a considerable speed reduction is required for long periods.

(ii) **Series-wound motors**

The speed of a d.c. series-wound motor is given by:

$$n = k\left(\frac{V-IR}{\Phi}\right)$$

where k is a constant, V is the terminal voltage, R is the combined resistance of the armature and series field and Φ is the flux.

Thus, a reduction in flux results in an increase in speed. This is achieved by putting a variable resistance in parallel with the field winding and reducing the field current, and hence flux, for a given value of supply current. A circuit diagram of this arrangement is shown in *Figure 22.10(a)*. A variable resistor

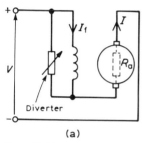 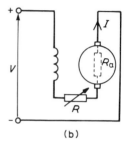

(a)   (b)

Figure 22.10

connected in parallel with the series-wound field to control speed is called a **diverter**. Speeds above those given with no diverter are obtained by this method.

Speeds below normal are obtained by connecting a variable resistor in series with the field winding and armature circuit, as shown in *Figure 22.10(b)*. This effectively increases the value of R in the equation

$$n = k\left(\frac{V - IR}{\Phi}\right)$$

and thus reduces the speed. Since the additional resistor carries the full supply current, a large power loss is associated with large motors in which a considerable speed reduction is required for long periods.

23 a.c. motors

1 Two principal types of a.c. motors are in widespread use, these being **induction motors** and **synchronous motors**. Each of these types may be further subdivided into those being run from a three-phase industrial supply, called 3-phase motors and those largely in domestic use, called single-phase motors.

Three-phase induction motors

2 In d.c. motors, introduced in Chapter 22, conductors on a rotating armature pass through a stationary magnetic field. In a three-phase induction motor, the magnetic field rotates and this has the advantage that no external electrical connections to the rotor need be made. The result is a motor which: (i) is cheap and robust, (ii) is explosion proof, due to the absence of a commutator or slip-rings and brushes with their associated sparking, (iii) requires little or no skilled maintenance, and (iv) has self starting properties when switched to a supply with no additional expenditure on auxiliary equipment. The principal disadvantage of a three-phase induction motor is that its speed cannot be readily adjusted.

3 **Production of a rotating magnetic field**. When a three-phase supply is connected to symmetrical three-phase stator windings, the currents flowing in the windings produce a magnetic field. This magnetic field is constant in magnitude and rotates at constant speed as shown below, and is called the **synchronous speed**.

With reference to *Figure 23.1*, the windings are represented by three single-loop conductors, one for each phase, marked $R_S R_F$, $Y_S Y_F$ and $B_S B_F$, the S and F signifying start and finish. In practice, each phase winding comprises many turns and is distributed around the stator; the single-loop approach is for clarity only.

When the stator windings are connected to a three-phase supply, the current flowing in each winding varies with time and is as shown in *Figure 23.1(a)*. If the value of current in a winding is positive, the assumption is made that it flows from start to finish of the winding, i.e. if it is the red-phase, current flows from R_S to R_F,

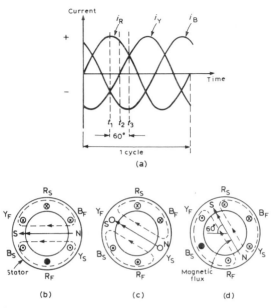

Figure 23.1

i.e. away from the viewer in R_S and towards the viewer in R_F. When the value of current is negative, the assumption is made that it flows from finish to start, i.e. towards the viewer in an 'S' winding and away from the viewer in an 'F' winding.

At time, say t_1, shown in *Figure 23.1(a)*, the current flowing in the red phase is a maximum possible value. At the same time, t_1, the currents flowing in the yellow and blue phases are both 0.5 times the maximum value and are negative. The current distribution in the stator windings is therefore as shown in *Figure 23.1(b)*, in which current flows away from the viewer (shown as X) in R_S since it is positive but towards the viewer (shown as .) in Y_S and B_S, since these are negative. The resulting magnetic field is as shown, due to the 'solenoid' action and application of the corkscrew rule.

A short time later at time t_2, the current flowing in the red phase has fallen to about 0.87 times its maximum value and is positive, the current in the yellow phase is zero and the current in

175

the blue phase is about 0.87 times its maximum value and is negative. Hence the currents and resultant magnetic field are as shown in *Figure 23.1(c)*. At time t_3, the currents in the red and yellow phases are 0.5 of their maximum values and the current in the blue phase is a maximum negative value. The currents and resultant magnetic field are shown in *Figure 23.1(d)*.

Similar diagrams to *Figure 23.1(b)*, *(c)*, *(d)* can be produced for all time values and these would show that the magnetic field travels through one revolution for each cycle of the supply voltage applied to the stator windings. By considering the flux values rather than the current values, it can be shown that the rotating magnetic field has a constant value of flux.

4 The rotating magnetic field produced by three phase windings could have been produced by rotating a permanent magnet's north and south pole at synchronous speed (shown as N and S at the ends of the flux phasors in *Figures 23.1(b)*, *(c)* and *(d)*). For this reason, it is called a 2-pole system and an induction motor using three phase windings only is called a 2-pole induction motor.

If six windings displaced from one another by 60° are used, as shown in *Figure 23.2(b)*, by drawing the current and resultant magnetic field diagrams at various time values, it may be shown that one cycle of the supply current to the stator windings causes the magnetic field to move through half a revolution. The current distribution in the stator windings is shown in *Figure 23.2(b)*, for the time t shown in *Figure 23.2(a)*.

It can be seen that for six windings on the stator, the magnetic flux produced is the same as that produced by rotating two permanent magnet north poles and two permanent magnet south poles at synchronous speed. This is called a 4-pole system and an induction motor using six phase windings is called a 4-pole induction motor. By increasing the number of phase windings the number of poles can be increased to any even number. In general, if f is the frequency of the currents in the stator windings and the stator is wound to be equivalent to p pairs of poles, the speed of revolution of the rotating magnetic field, i.e. the synchronous speed, n_s is given by:

$$n_s = \frac{f}{p} \text{ rev/s.}$$

5 **The principle of operation of the three-phase induction motor**

The stator of a three-phase induction motor is the stationary part corresponding to the yoke of a d.c. machine. It is wound to

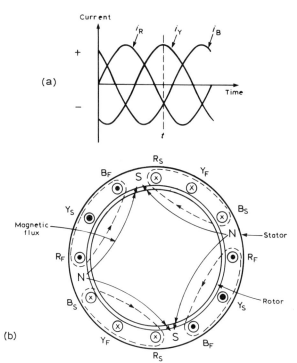

Figure 23.2

give a 2-pole, 4-pole, 6-pole, ... rotating magnetic field, depending on the rotor speed required. The rotor, corresponding to the armature of a d.c. machine, is built up of laminated iron, to reduce eddy currents.

In the type most widely used, known as a squirrel-cage rotor, copper or aluminium bars are placed in slots cut in the laminated iron, the ends of the bars being welded or brazed into a heavy conducting ring, (see *Figure 23.3(a)*). A cross-sectional view of a three-phase induction motor is shown in *Figure 23.3(b)*.

When a three-phase supply is connected to the stator windings, a rotating magnetic field is produced. As the magnetic flux cuts a bar on the rotor, an e.m.f. is induced in it and since it is joined, via the end conducting rings, to another bar one pole

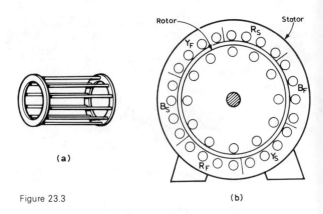

Figure 23.3

pitch away, a current flows in the bars. The magnetic field associated with this current flowing in the bars interacts with the rotating magnetic field and a force is produced, tending to turn the rotor in the same direction as the rotating magnetic field (see *Figure 23.4*).

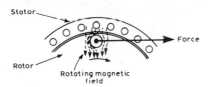

Figure 23.4

6 **Slip**

The force exerted by the rotor bars causes the rotor to turn in the direction of the rotating magnetic field. As the rotor speed increases, the rate at which the rotating magnetic field cuts the rotor bars is less and the frequency of the induced e.m.f.'s in the rotor bars is less. If the rotor runs at the same speed as the rotating magnetic field, no e.m.f.'s are induced in the rotor, hence there is no force on them and no torque on the rotor. Thus the rotor slows down. For this reason the rotor can never run at synchronous speed.

When there is no load on the rotor, the resistive forces due to windage and bearing friction are small and the rotor runs very nearly at synchronous speed. As the rotor is loaded, the speed falls and this causes an increase in the frequency of the induced e.m.f.'s in the rotor bars and hence the rotor current, force and torque increase. The difference between the rotor speed, n_r, and the synchronous speed, n_s, is called the slip speed, i.e.

slip speed $= n_s - n_r$ rev/s

The ratio $(n_s - n_r)/n_s$ is called the fractional slip or just the slip, s, and is usually expressed as a percentage. Thus

$$\text{slip, } s = \frac{n_s - n_r}{n_s} \times 100\%.$$

Typical values of slip between no load and full load are about 4% to 5% for small motors and $1\frac{1}{2}$% to 2% for large motors.

7   Three-phase induction motors are widely used in industry and constitute almost all industrial drives where a nearly constant speed is required, from small workshops to the largest industrial enterprises.

Single-phase induction motor

8   The majority of 'fractional horse-power' motors in domestic use for driving refrigerators, hot water pumps, fans, hair dryers and so on are single-phase induction motors. A single-phase supply connected to the stator winding only produces a pulsating magnetic field rather than the rotating magnetic field of a three-phase supply and facilities such as additional windings have to be built-in to the stators of these motors to give some torque at zero speed. There are several devices used and motors are named after the particular device used. Some of these include:
split-phase start,
capacitor start,
capacitor start and run,
shaded pole,
permanent-split capacitor induction motors, and so on.

9   **Three-phase synchronous motors** normally have two windings, a three-phase stator winding to produce a rotating magnetic field and a rotor winding supplied by a direct current to magnetise the rotor. The rotor is locked magnetically to the rotating magnetic field and for a constant frequency supply, the rotor runs at a constant speed which is directly proportional to the supply frequency.

These motors may be used on many of the loads driven by induction motors but (i) are normally more expensive, (ii) require both a.c. and d.c. supplies, (iii) require additional equipment to run them up to near to their normal running speed, so that the rotor magnetic field can be locked to the rotating magnetic field, and (iv) have higher maintenance costs than induction motors. However, they can run at higher efficiencies and when used adjacent to several induction motors can lead to a higher efficiency of the whole system (power factor correction). Due to the disadvantages listed above, three-phase synchronous motors are normally used for applications requiring a large power input, where savings due to higher efficiency outweigh the disadvantages. Their uses include driving large water pumps (power stations, water supply), driving rolling lines in steel mills and driving mine ventilating fans.

10   As for single-phase induction motors, **single-phase synchronous motors** require additional starting devices. They also require a direct current supply to the rotor or a permanent magnet built into the rotor. The former is rarely used and the principal uses of single-phase synchronous motors are largely limited to electric clocks, timing devices, record players and so on, these motors being called 'shaded-pole motors'. As for three-phase synchronous motors, they run at constant speed when connected to a constant frequency supply.

24 Electrical measuring instruments and measurements

1   Tests and measurements are important in designing, evaluating, maintaining and servicing electrical circuits and equipment. In order to detect electrical quantities such as current, voltage, resistance or power, it is necessary to transform an electrical quantity or condition into a visible indication. This is done with the aid of instruments (or meters) that indicate the magnitude of quantities either by the position of a pointer moving over a graduated scale (called an analogue instrument) or in the form of a decimal number (called a digital instrument).

2   All analogue electrical indicating instruments require three essential devices. These are:

   (a)   **A deflecting or operating device** A mechanical force is produced by the current or voltage which causes the pointer to deflect from its zero position.
   (b)   **A controlling device** The controlling force acts in opposition to the deflecting force and ensures that the deflection shown on the meter is always the same for a given measured quantity. It also prevents the pointer always going to the maximum deflection. There are two main types of controlling device — spring control and gravity control.
   (c)   **A damping device** The damping force ensures that the pointer comes to rest in its final position quickly and without undue oscillation. There are three main types of damping used — eddy-current damping, air-friction damping and fluid-friction damping.

3   There are basically two types of scale — linear and non-linear. A **linear scale** is shown in *Figure 24.1(a)* where each of the divisions or graduations are evenly spaced. The voltmeter shown has a range 0–100 V, i.e. a full scale deflection (FSD) of 100 V.

   A **non-linear scale** is shown in *Figure 24.1(b)*. The scale is cramped at the beginning and the graduations are uneven throughout the range. The ammeter shown has a FSD of 10 A.

4 Comparison of the moving coil, moving iron and moving coil rectifier instruments

Type of instrument	Moving coil	Moving iron	Moving coil rectifier
Suitable for measuring:	Direct current and voltage.	Direct and alternating current and voltage (reading in r.m.s. value).	Alternating current and voltage (reads average value but scale is adjusted to give r.m.s. value for sinusoidal waveforms).
Scale	Linear	Non-linear	Linear
Method of control	Hairsprings	Hairsprings	Hairsprings
Method of damping	Eddy current	Air	Eddy current
Frequency limits	—	20 Hz–200 Hz	20 Hz–100 kHz

Advantages	1. Linear scale. 2. High sensitivity. 3. Well shielded from stray magnetic fields. 4. Lower power consumption.	1. Robust construction. 2. Relatively cheap. 3. Measures d.c. and a.c. 4. In frequency range 20–100 Hz reads r.m.s. correctly, regardless of supply waveform.	1. Linear scale. 2. High sensitivity. 3. Well shielded from stray magnetic fields. 4. Lower power consumption. 5. Good frequency range.
Disadvantages	1. Only suitable for d.c. 2. More expensive than moving iron type. 3. Easily damaged.	1. Non-linear scale. 2. Affected by stray magnetic fields. 3. Hysteresis errors in d.c. circuits. 4. Liable to temperature errors. 5. Due to the inductance of the solenoid, readings can be affected by variation of frequency.	1. More expensive than moving iron type. 2. Errors caused when supply is non-sinusoidal.

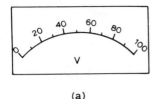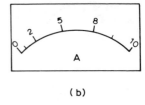

(a)   (b)

Figure 24.1

Principle of operation of a moving coil instrument

5   A moving coil instrument operates on the motor principle. When a conductor carrying current is placed in a magnetic field, a force F is exerted on the conductor, given by $F = BIl$. If the flux density B is made constant (by using permanent magnets) and the conductor is a fixed length (say, a coil) then the force will depend only on the current flowing in the conductor.

In a moving coil instrument, a coil is placed centrally in the gap between shaped pole pieces as shown by the front elevation in *Figure 24.2(a)*. The coil is supported by steel pivots, resting in jewel bearings, on a cylindrical iron core. Current is led into and out of the coil by two phosphor bronze spiral hairsprings which are wound in opposite directions to minimise the effect of temperature change and to limit the coil swing (i.e. to control the movement) and return the movement to zero position when no current flows.

Current flowing in the coil produces forces as shown in *Figure 24.2(b)*, the directions being obtained by Fleming's left-hand rule. The two forces, F_A and F_B, produce a torque which will move the coil in a clockwise direction, i.e. move the pointer from left to right. Since force is proportional to current the scale is linear.

When the aluminium frame, on which the coil is wound, is rotated between the poles of the magnet, small currents (called eddy currents) are induced into the frame, and this provides automatically the necessary damping of the system due to the reluctance of the former to move within the magnetic field.

The moving coil instrument will only measure direct current or voltage and the terminals are marked positive and negative to ensure that the current passes through the coil in the correct direction to deflect the pointer 'up the scale'.

The range of this sensitive instrument is extended by using shunts and multipliers. (See para. 10.)

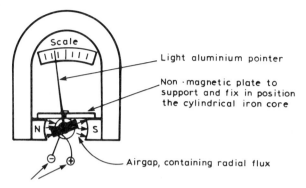

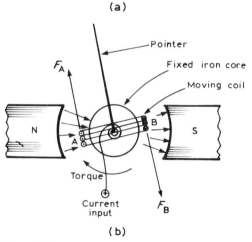

Figure 24.2

Principle of operation of the moving iron instrument

6   (a) An **attraction type** of moving iron instrument is shown
        diagrammatically in *Figure 24.3(a)*. When current flows
        in the solenoid, a pivoted soft iron disc is attracted

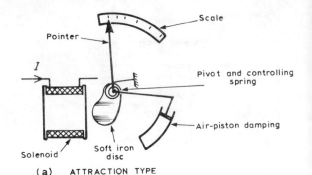

(a) ATTRACTION TYPE

(b) REPULSION TYPE

Figure 24.3

towards the solenoid and the movement causes a pointer to move across a scale.

(b) In the **repulsion type** moving iron instrument shown diagrammatically in *Figure 24.3(b)*, two pieces of iron are placed inside the solenoid, one being fixed, and the other attached to the spindle carrying the pointer.

When current passes through the solenoid, the two pieces of iron are magnetised in the same direction and therefore repel each other. The pointer thus moves across the scale. The force moving the pointer is, in each type, proportional to I^2. Because of this the direction of current does not matter and the moving iron instrument can be used on d.c. or a.c. The scale, however, is non-linear.

7 A moving coil instrument, which measures only d.c., may be used in conjunction with a bridge rectifier circuit as shown in *Figure 24.4* to provide an indication of alternating currents and voltages. The average value of the full wave rectified current is

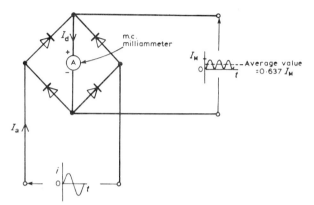

Figure 24.4

$0.637I_M$. However, a meter being used to measure a.c. is usually calibrated in r.m.s. values. For sinusoidal quantities the indication is

$$\frac{0.707I_M}{0.637I_M}, \text{ i.e. 1.11 times the mean value.}$$

Rectifier instruments have scales calibrated in r.m.s. quantities and it is assumed by the manufacturer that the a.c. is sinusoidal.

8 An **ammeter**, which measures current, has a low resistance (ideally zero) and must be connected in series with the circuit.

9 A **voltmeter**, which measures p.d., has a high resistance (ideally infinite) and must be connected in parallel with the part of the circuit whose p.d. is required.

Shunts and multipliers

10 There is no difference between the basic instrument used to measure current and voltage since both use a milliammeter as their

basic part. This is a sensitive instrument which gives FSD for currents of only a few milliamperes. When an ammeter is required to measure currents of larger magnitude, a proportion of the current is diverted through a low value resistance connected in parallel with the meter. Such a diverting resistance is called a **shunt**.

From *Figure 24.5(a)*, $V_{PQ} = V_{RS}$.

Hence, $I_a r_a = I_S R_S$

Thus the value of the shunt, $R_S = \dfrac{I_a r_a}{I_S}$ Ω.

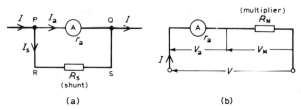

(a)   (b)

Figure 24.5

The milliammeter is converted into a voltmeter by connecting a high resistance (called a **multiplier**) in series with it as shown in *Figure 24.5(b)*. From *Figure 24.5(b)*, $V = V_a + V_M = Ir_a + IR_M$

Thus the value of the multiplier, $R_M = \dfrac{V - Ir_a}{I}$ Ω.

For example, let a m.c. instrument have a FSD of 20 mA and a resistance of 25 Ω. To enable the instrument to be used as a 0–10 A ammeter, a shunt resistance R_S needs to be connected in parallel with the instrument. From *Figure 24.5(a)*,

$I = 10$ A, $I_S = I - I_a = 10 - 0.020 = 9.98$ A.

Hence the value of R_S is given by:

$$R_S = \frac{I_a r_a}{I_S} = \frac{(0.020)(25)}{9.98} = \textbf{50.10 m}\Omega.$$

To enable the instrument to be used as a 0 to 100 V voltmeter, a multiplier R_M needs to be connected in series with the instrument, the value of R_M being given by:

$$R_M = \frac{V - Ir_a}{I} = \frac{100 - (0.020)(25)}{0.020} = \textbf{4.975 k}\Omega.$$

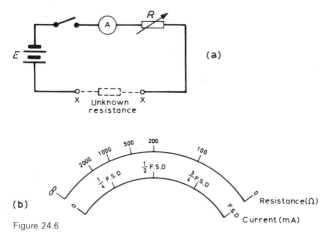

Figure 24.6

11 An **ohmmeter** is an instrument for measuring electrical
resistance. A simple ohmmeter circuit is shown in *Figure 24.6(a)*.
Unlike the ammeter or voltmeter, the ohmmeter circuit does not
receive the energy necessary for its operation from the circuit under
test. In the ohmmeter this energy is supplied by a self-contained
source of voltage, such as a battery. Initially, terminals XX are
short-circuited and R adjusted to give FSD on the milliammeter. If
current I is at a maximum value and voltage E is constant, then
resistance $R=E/I$ is at a minimum value. Thus FSD on the
milliammeter is made zero on the resistance scale. When terminals
XX are open circuited no current flows and R $(=E/O)$ is infinity,
∞. The milliammeter can thus be calibrated directly in ohms. A
cramped (non-linear) scale results and is 'back to front' (as shown
in *Figure 24.6(b)*). When calibrated, an unknown resistance is
placed between terminals XX and its value determined from the
position of the pointer on the scale. An ohmmeter designed for
measuring low values of resistance is called a **continuity tester**.
An ohmmeter designed for measuring high values of resistance (i.e.
megohms) is called an **insulation resistance tester** (or
'**megger**').
12 Instruments are manufactured that combine a moving coil
meter with a number of shunts and series multipliers, to provide a
range of readings on a single scale graduated to read current and
voltage. If a battery is incorporated into the instrument then
resistance can also be measured.

Such instruments are called **multimeters** or **universal instruments** or **multirange instruments**. An **Avometer** is a typical example. A particular range may be selected either by the use of separate terminals or by a selector switch. Only one measurement can be performed at one time. Often such instruments can be used in a.c. as well as d.c. circuits when a rectifier is incorporated in the instrument.

13 A **wattmeter** is an instrument for measuring electrical power in a circuit. *Figure 24.7* shows typical connections of a wattmeter used for measuring power supplied to a load. The instrument has two coils:

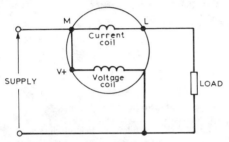

Figure 24.7

(i) a **current coil**, which is connected in series with the load (like an ammeter), and
(ii) a **voltage coil**, which is connected in parallel with the load (like a voltmeter).

14 **The cathode ray oscilloscope (CRO)** may be used in the observation of waveforms and for the measurement of voltage, current, frequency, phase and periodic time.

(i) With **direct voltage measurements**, only the Y amplifier 'volts/cm' switch on the CRO is used. With no voltage applied to the Y plates the position of the spot trace on the screen is noted. When a direct voltage is applied to the Y plates the new position of the spot trace is an indication of the magnitude of the voltage. For example, in *Figure 24.8(a)*, with no voltage applied to the Y plates, the spot trace is in the centre of the screen (initial position) and then the spot trace moves 2.5 cm to the final position shown, on application of a d.c. voltage. With the 'V/cm' switch on 10 V/cm the magnitude of the direct voltage is 2.5 cm × 10 V/cm, i.e. 25 V.

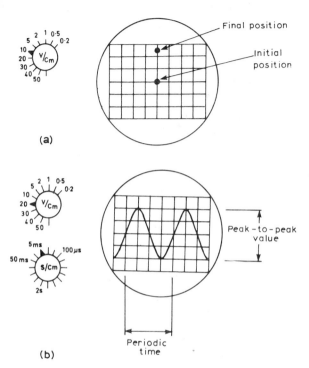

(a)

(b)

Figure 24.8

(ii) With **alternating voltage measurements**, let a sinusoidal waveform be displayed on a CRO screen as shown in *Figure 24.8(b)*. If the 'variable' switch is on, say, 5 ms/cm then the **periodic time** T of the sine wave is 5 ms/cm × 4 cm, i.e. 20 ms or 0.02 s.

Since frequency $f = \dfrac{1}{T}$,

$$\textbf{frequency} = \frac{1}{0.02} = \textbf{50 Hz}.$$

If the 'volts/cm' switch is on, say, 20 V/cm then the **amplitude** or

peak value of the sine wave shown is 20 V/cm × 2 cm, i.e. 40 V.

Since r.m.s. voltage = $\dfrac{\text{peak voltage}}{\sqrt{2}}$

r.m.s. voltage $= \dfrac{40}{\sqrt{2}} = \mathbf{28.28\ V}$.

Double beam oscilloscopes are useful whenever two signals are to be compared simultaneously. The CRO demands reasonable skill in adjustment and use. However its greatest advantage is in observing the shape of a waveform — a feature not possessed by other measuring instruments.

15 An **electronic voltmeter** can be used to measure with accuracy e.m.f. or p.d. from millivolts to kilovolts by incorporating in its design amplifiers and attentuators.

16 A **null method of measurement** is a simple, accurate and widely used method which depends on an instrument reading being adjusted to read zero current only. The method assumes:
(i) if there is any deflection at all, then some current is flowing, and
(ii) if there is no deflection, then no current flows (i.e. a null condition).

Hence it is unnecessary for a meter sensing current flow to be calibrated when used in this way. A sensitive milliammeter or microammeter with centre zero position setting is called a **galvanometer**. Two examples where the method is used are in the Wheatstone bridge and in the d.c. potentiometer.

17 A **Wheatstone bridge**, shown in *Figure 24.9*, is used in d.c. circuits to compare an unknown resistance R_x with others of known values. R_3 is varied until zero deflection is obtained on the galvanometer, G. At balance (i.e. zero deflection on the galvanometer) the products of diagonally opposite resistances are equal to one another,

i.e. $R_1 R_x = R_2 R_3$

from which, $\boxed{R_x = \dfrac{R_2 R_3}{R_1}\ \textbf{ohms}.}$

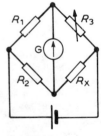

Figure 24.9

18 The **d.c. potentiometer** is a null-balance instrument used for determining values of e.m.f.'s and p.d.'s by comparison with a known e.m.f. or p.d. In *Figure 24.10(a)*, using a standard cell of known e.m.f. E_1, the slider S is moved along the slide wire until balance is obtained (i.e. the galvanometer deflection is zero),

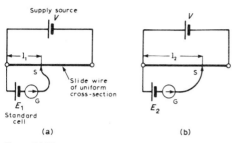

Figure 24.10

shown as length l_1. The standard cell is now replaced by a cell of unknown e.m.f., E_2 (see *Figure 24.10(b)*) and again balance is obtained (shown as l_2). Since $E_1 \propto l_1$ and $E_2 \propto l_2$,

then $\dfrac{E_1}{E_2} = \dfrac{l_1}{l_2}$ and $\boxed{E_2 = E_1\left(\dfrac{l_2}{l_1}\right) \textbf{ volts.}}$

A.C. bridges

19 A Wheatstone bridge type circuit, shown in *Figure 24.11*, may be used in a.c. circuits to determine unknown values of inductance and capacitance, as well as resistance. When the potential differences across Z_3 and Z_x (or across Z_1 and Z_2) are equal in magnitude and phase, then the current flowing through the galvanometer, G, is zero. At balance, $Z_1Z_x = Z_2Z_3$, from which

$$Z_x = \frac{Z_2 Z_3}{Z_1} \ \Omega$$

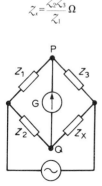

Figure 24.11

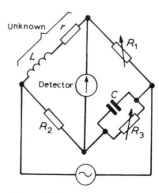

Figure 24.12

20   There are many forms of a.c. bridge, and these include: the Maxwell, Hay, Owen and Heaviside bridges for measuring inductance, and the De Sauty, Schering and Wien bridges for measuring capacitance. A commercial or universal bridge is one which can be used to measure resistance, inductance or capacitance.

21   **Maxwell's bridge** for measuring the inductance L and resistance r of an inductor is shown in *Figure 24.12*.

At balance, $Z_1 Z_2 = Z_3 Z_4$

Using complex quantities,

$$Z_1 = R_1, \ Z_2 = R_2$$

$$Z_3 = \frac{R_3(-jX_C)}{(R_3 - jX_C)} \left(\text{i.e. } \frac{\text{product}}{\text{sum}} \right) \text{ and } Z_4 = r + jX_L$$

Hence, $R_1 R_2 = \dfrac{R_3(-jX_C)}{(R_3 - jX_C)} \ (r + jX_L)$

i.e. $R_1 R_2 (R_3 - jX_C) = (-jR_3 R_C)(r + jX_L)$

$R_1 R_2 R_3 - jR_1 R_2 X_C = -jrR_3 X_C - j^2 R_3 X_C X_L$

i.e. $R_1 R_2 R_3 - jR_1 R_2 X_C = -jrR_3 X_C + R_3 X_C X_L$ (since $j^2 = -1$)

Equating the real parts gives: $R_1 R_2 R_3 = R_3 X_C X_L$

from which, $X_L = \dfrac{R_1 R_2}{X_C}$

i.e. $2\pi f L = \dfrac{R_1 R_2}{\dfrac{1}{2\pi f C}} = R_1 R_2 (2\pi f C)$

Hence **inductance**, $L = R_1 R_2 C$ **henry** $\qquad\qquad (1)$

Equating the imaginary parts gives:

$$-R_1 R_2 X_C = -rR_3 X_C$$

from which, **resistance** $r = \dfrac{R_1 R_2}{R_3} \ \Omega.$ $\qquad\qquad (2)$

From equation (1), $R_2 = \dfrac{L}{R_1 C}$ and from

equation (2), $R_3 = \dfrac{R_1}{r} R_2$. Hence $R_3 = \dfrac{R_1}{r}\left(\dfrac{L}{R_1 C}\right) = \dfrac{L}{Cr}$

If the frequency is constant then

$$R \propto \frac{L}{r} \propto \frac{L}{r} w \propto Q\text{-factor.}$$

Thus the bridge can be adjusted to give a direct indication of Q-factor.

22   The **Q-factor** for a series L-C-R circuit is the voltage magnification at resonance,

i.e. Q-factor$= \dfrac{\text{voltage across capacitor}}{\text{supply voltage}} = \dfrac{V_c}{V}$

(see Chapter 14, para. 15).

The simplified circuit of a **Q-meter**, used for measuring Q-factor, is shown in *Figure 24.13*. Current from a variable frequency

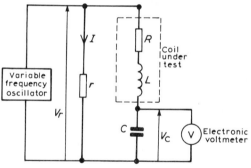

Figure 24.13

oscillator flowing through a very low resistance r develops a variable frequency voltage, V_r, which is applied to a series L-R-C circuit. The frequency is then varied until resonance causes voltage V_c to reach a maximum value. At resonance V_r and V_c are noted.

Then Q-factor$= \dfrac{V_c}{V_r} = \dfrac{V_c}{Ir}$

In a practical Q-meter, V_r is maintained constant and the electronic voltmeter can be calibrated to indicate the Q-factor directly. If a variable capacitor C is used and the oscillator is set at a given frequency, then C can be adjusted to give resonance. In this way inductance L may be calculated using

$$f_r = \frac{1}{2\pi\sqrt{(LC)}}$$

Since $Q = \frac{2\pi f L}{R}$, then R may be calculated.

Q-meters operate at various frequencies and instruments exist with frequency ranges from 1 kHz to 50 MHz. Errors in measurement can exist with Q-meters since the coil has an effective parallel self capacitance due to capacitance between turns. The accuracy of a Q-meter is approximately $\pm 5\%$.

Waveform harmonics

23 (i) Let an instantaneous voltage v be represented by $v = V_M \sin 2\pi f t$ volts. This is a waveform which varies sinusoidally with time t, has a frequency f, and a maximum value, V_M. Alternating voltages are usually assumed to have wave-shapes which are sinusoidal where only one frequency is present. If the wave-form is not sinusoidal it is called a **complex wave**, and, whatever its shape, it may be split up mathematically into components called the **fundamental** and a number of **harmonics**. This process is called **harmonic analysis**. The fundamental (or first harmonic) is sinusoidal and has the supply frequency, f, the other harmonics are also sine waves having frequencies which are integer multiples of f. Thus, if the supply frequency is 50 Hz, then the third harmonic frequency is 150 Hz, the fifth 250 Hz, and so on.

(ii) A complex waveform comprising the sum of the fundamental and a third harmonic of about half the amplitude of the fundamental is shown in *Figure 24.14(a)*, both waveforms being initially in phase with each other. If further odd harmonic waveforms of the appropriate amplitudes are added, a good approximation to a square wave results.

In *Figure 24.14(b)* the third harmonic is shown having an initial phase displacement from the fundamental. The positive and negative half cycles of each of the complex waveforms shown in *Figure 24.14(a)* and *(b)* are identical in shape, and this is a feature of waveforms containing the fundamental and only odd harmonics.

(iii) A complex waveform comprising the sum of the fundamental and a second harmonic of about half the amplitude of the fundamental is shown in *Figure 24.14(c)*, each waveform being initially in phase with each other. If further even harmonics of appropriate amplitudes are added a good approximation to a

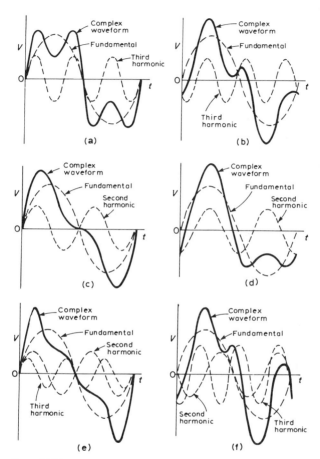

Figure 24.14

triangular wave results. In *Figure 24.14(c)* the negative cycle appears as a mirror image of the positive cycle about point A. In *Figure 24.14(d)* the second harmonic is shown with an initial phase displacement from the fundamental and the positive and negative half cycles are dissimilar.

(iv) A complex waveform comprising the sum of the fundamental, a second harmonic and a third harmonic is shown in *Figure 24.14(e)*, each waveform being initially 'in-phase'. The negative half cycle appears as a mirror image of the positive cycle about point B. In *Figure 24.14(f)*, a complex waveform comprising the sum of the fundamental, a second harmonic and a third harmonic are shown with initial phase displacement. The positive and negative half cycles are seen to be dissimilar.

The features mentioned relative to *Figures 24.14(a)* to *(f)* make it possible to recognise the harmonics present in a complex waveform displayed on a CRO.

24   Some measuring instruments depend for their operation on power taken from the circuit in which measurements are being made. Depending on the 'loading' effect of the instrument (i.e. the current taken to enable it to operate), the prevailing circuit conditions may change. The resistance of voltmeters may be calculated since each have a stated **sensitivity** (or '**figure of merit**'), often stated in 'kΩ per volt' of FSD. A voltmeter should have as high a resistance as possible ($-$ ideally infinite). In a.c. circuits the impedance of the instrument varies with frequency and thus the loading effect of the instrument can change. Electronic measuring instruments have advantages over instruments such as the moving iron or moving coil meters, in that they have a much higher input resistance (some as high as 1000 MΩ) and can handle a much wider range of frequency (from d.c. up to MHz).

25   Instruments for a.c. measurements are generally calibrated with a sinusoidal alternating waveform to indicate r.m.s. values when a sinusoidal signal is applied to the instrument. Some instruments, such as the moving iron and electrodynamic instruments, give a true r.m.s. indication. With other instruments the indication is either scaled up from the mean value (such as with the rectifier moving coil instrument) or scaled down from the peak value. Sometimes quantities to be measured have complex waveforms (see para. 23), and whenever a quantity is non-sinusoidal, errors in instrument readings can occur if the instrument has been calibrated for sine waves only. Such waveform errors can be largely eliminated by using electronic instruments.

26   **Errors** are always introduced when using instruments to measure electrical quantities. Besides possible errors introduced by the operator or by the instrument disturbing the circuit, errors are caused by the limitations of the instrument used.

The calibration accuracy of an instrument depends on the precision with which it is constructed. Every instrument has a margin of error which is expressed as a percentage of the instruments full scale deflection. For example, an instrument may have an accuracy of $\pm 2\%$ of FSD. Thus, if a voltmeter has a FSD of 100 V and it indicates say 60 V, then the actual voltage measured may be anywhere between 60 V $\pm$ (2% of 100 V), i.e. 60 ± 2 V, i.e. between 58 V and 62 V. As a percentage of the voltmeter reading this error is $\pm 2/60 \times 100\%$, i.e. $\pm 3.33\%$. Hence the accuracy can be expressed as 60 V $\pm 3.33\%$. It follows that an instrument having a 2% FSD accuracy can give relatively large errors when operating at conditions well below FSD.

When more than one instrument is used in a circuit then a cumulative error results. For example, if the current flowing through and the p.d. across a resistor is measured, then the percentage error in the ammeter is added to the percentage error in the voltmeter when determining the maximum possible error in the measured value of resistance.

Decibel units

27   In electronic systems, the ratio of two similar quantities measured at different points in the system, are often expressed in logarithmic units. By definition, if the ratio of two powers P_1 and P_2 is to be expressed in decibel (dB) units then the number of decibels, X, is given by:

$$X = 10 \lg \left(\frac{P_2}{P_1} \right) \mathbf{dB} \tag{1}$$

Thus, when the power ratio, $\frac{P_2}{P_1} = 1$,

then the decibel power ratio $= 10 \lg 1 = 0$

when the power ratio, $\frac{P_2}{P_1} = 100$,

then the decibel power ratio $= 10 \lg 100 = +20$

(i.e. a power gain),

and when the power ratio, $\frac{P_2}{P_1} = \frac{1}{10}$,

then the decibel power ratio $= 10 \lg \frac{1}{10} = -10$,

(i.e. a power loss or attentuation).

28   Logarithmic units may also be used for voltage and current ratios. Power P, is given by $P = I^2 R$ or $P = V^2/R$.

Substituting in equation (1) gives:

$$X = 10 \lg \left(\frac{I_2^2 R_2}{I_1^2 R_1}\right) \text{ dB} \quad \text{or} \quad X = 10 \lg \left(\frac{V_2^2/R_2}{V_1^2/R_1}\right) \text{ dB}$$

If $R_1 = R_2$, then $X = 10 \lg \left(\frac{I_2^2}{I_1^2}\right)$ dB or $X = 10 \lg \left(\frac{V_2^2}{V_1^2}\right)$ dB

$$X = 20 \lg \left(\frac{I_2}{I_1}\right) \text{ dB} \quad \text{or} \quad X = 20 \lg \left(\frac{V_2}{V_1}\right) \text{ dB}$$

(from the laws of logarithms).

Thus if the current input to a system is 5 mA and the current output is 20 mA, the decibel current ratio

$$= 20 \lg \left(\frac{I_2}{I_1}\right) = 20 \lg \left(\frac{20}{5}\right) = 20 \lg 4 = \textbf{12 dB gain}.$$

29 From equation (1), X decibels is a logarithmic ratio of two similar quantities and is not an absolute unit of measurement. It is therefore necessary to state a **reference level** to measure a number of decibels above or below that reference. The most widely used reference level for power is 1 mW, and when power levels are expressed in decibels, above or below the 1 mW reference level, the unit given to the new power level is dBm.

A voltmeter can be re-scaled to indicate the power level directly in decibels. The scale is generally calibrated by taking a reference level of 0 dB when a power of 1 mW is dissipated in a 600 Ω resistor (this being the natural impedance of a simple transmission line). The reference voltage V is then obtained from

$$P = \frac{V^2}{R}, \text{ i.e. } 1 \times 10^{-3} = \frac{V^2}{600} \text{ from which, } V = 0.775 \text{ volts.}$$

In general, the number of dBm, $X = 20 \lg \left(\frac{V}{0.775}\right)$.

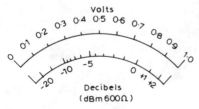

Figure 24.15

Thus $V = 0.20$ V corresponds to $20 \lg \left(\dfrac{0.20}{0.775} \right) = -11.77$ dBm and

$V = 0.90$ V corresponds to $20 \lg \left(\dfrac{0.90}{0.775} \right) = +1.3$ dBm, and so on.

A typical **decibelmeter**, or **dB meter scale** is shown in *Figure 24.15*. Errors are introduced with dB meters when the circuit impedance is not 600 Ω.

25 Speed and velocity

1 Speed is the rate of covering distance and is given by:

$$\text{speed} = \frac{\textbf{distance travelled}}{\textbf{time taken}}$$

The usual units for speed are metres per second (m/s or m s^{-1}), or kilometres per hour (km/h or km h^{-1}). Thus if a person walks 5 kilometres in 1 hour, the speed of the person is $\frac{5}{1}$, that is, 5 kilometres per hour.

The symbol for the SI unit of speed and velocity is written as 'm s^{-1}', called the 'index notation'. However, engineers usually use the symbol m/s, called the 'oblique notation', and it is this notation which is largely used in this chapter and other chapters on mechanics. One of the exceptions is when labelling the axes of graphs, when two obliques occur, and in this case the index notation is used. Thus for speed or velocity, the axis markings are speed/m s^{-1} or velocity/m s^{-1}.

2 One way of giving data on the motion of an object is graphically. A graph of distance travelled (the scale on the vertical axis of the graph), against time (the scale on the horizontal axis of the graph), is called a **distance-time graph**. Thus if a plane travels 500 km in its first hour of flight and 750 km in its second hour of flight, then after 2 h, the total distance travelled is (500 + 750) kilometres, that is, 1250 km. The distance-time graph for this flight is shown in *Figure 25.1*.

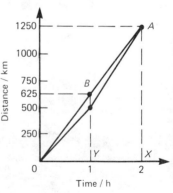

Figure 25.1

202

3   The **average speed** is given by

total distance travelled

total time taken

Thus, the average speed of the plane in para. 2 is:

$$\frac{(500+750) \text{ km}}{(1+1) \text{ h}}, \text{ i.e. } \frac{1250}{2} \text{ or } 625 \text{ km/h}.$$

If points O and A are joined in *Figure 25.1*, the slope of line OA is defined as

change in distance (vertical)

change in time (horizontal)

for any two points on line OA.

For point A, the change in distance is AX, that is 1250 km, and the change in time is OX, that is, 2 h. Hence the average speed is

$$\frac{1250}{2}, \text{ i.e. } 625 \text{ km/h}$$

Alternatively, for point B on line OA, the change in distance is BY, that is, 625 km and the change in time is OY, that is, 1 h, hence the average speed is

$$\frac{625}{1}, \text{ that is, } 625 \text{ km/h}$$

In general, the average speed of an object travelling between points M and N is given by the slope of line MN on the distance-time graph.

4   The **velocity** of an object is the speed of the object **in a specified direction**. Thus, if a plane is flying due south at 500 km/h, its speed is 500 km/h, but its velocity is 500 km/h **due south**. It follows that if the plane had flown in a circular path for one hour at a speed of 500 km/h hour, so that one hour after taking off it is again over the airport, its average velocity in the first hour of flight is zero.

5   The **average velocity** is given by:

distance travelled in a specific direction

time taken

If a plane flies from place O to place A, a distance of 300 km in 1 h, A being due north of O, then OA in *Figure 25.2* represents the

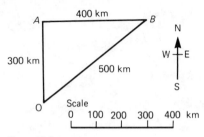

Figure 25.2

first hour of flight. It then flies from A to B, a distance of 400 km during the second hour of flight, B being due east of A, thus AB in *Figure 25.2* represents its second hour of flight. Its average velocity for the two hour flight is

$$\frac{\text{distance OB}}{2 \text{ hours}}, \text{ that is, } \frac{500 \text{ km}}{2 \text{ h}}$$

or 250 km/h in direction OB.

6   A graph of velocity (scale on the vertical axis), against time, (scale on the horizontal axis), is called **a velocity-time graph**. The graph shown in *Figure 25.3* represents a plane flying for 3 h at a constant speed of 600 km/h in a specified direction. The shaded area represents velocity, (vertically), multiplied by time (horizontally), and has units of kilometres/hours× hours, i.e. kilometres, and represents the distance travelled in a specific direction.

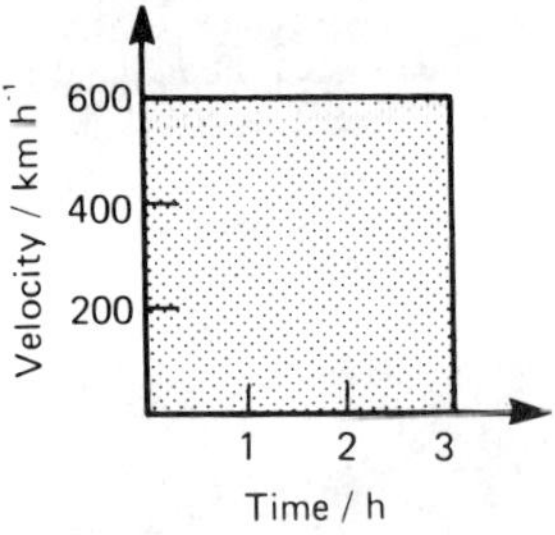

Figure 25.3

Another method of determining the distance travelled is from:

distance travelled = average velocity × time

Thus if a plane travels due south at 600 km/h for 20 minutes, the distance covered is

$$\frac{600 \text{ km}}{1 \text{ h}} \times \frac{20}{60} \text{ h, that is, 200 km.}$$

26 Acceleration and force

1   **Acceleration** is the rate of change of speed or velocity with time. The average acceleration, a, is given by:

$$a = \frac{\textbf{change in velocity}}{\textbf{time taken}}$$

The usual units are metres per second squared (m/s^2 or $m\ s^{-2}$). If u is the initial velocity of an object in m/s, v is the final velocity in m/s and t is the time in seconds elapsing between the velocities of u and v, then

average acceleration, $a = \dfrac{v - u}{t}$ m/s^2.

2   A graph of speed (scale on the vertical axis), against time (scale on the horizontal axis) is called a **speed-time graph**. For the speed-time graph shown in *Figure 26.1*, the slope of line OA is given by AX/OX. AX is the change in velocity from an initial

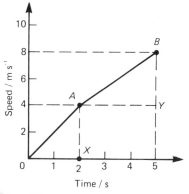

Figure 26.1

velocity u of zero to a final velocity, v, of 4 metres per second. OX is the time taken for this change in velocity, thus

$$\frac{AX}{OX} = \frac{\text{change in velocity}}{\text{time taken}}$$

$$= \text{the acceleration in the first two seconds.}$$

From the graph:

$$\frac{AX}{OX} = \frac{4 \text{ m/s}}{2 \text{ s}} = 2 \text{ m/s}^2,$$

i.e. the acceleration is 2 m/s^2.

Similarly, the slope of line AB in *Figure 26.1* is given by BY/AY, that is, the acceleration between 2 and 5 s is

$$\frac{8-4}{5-2} = \frac{4}{3} = 1\tfrac{1}{3} \text{ m/s}^2$$

In general, the slope of a line on a speed-time graph gives the acceleration.

3   If a dense object such as a stone is dropped from a height, called **free fall**, it has a constant acceleration of approximately 9.8 metres per second squared. In a vacuum, all objects have this same constant acceleration vertically downwards, that is, a feather has the same acceleration as a stone. However, if free fall takes place in air, dense objects have the approximately constant acceleration of 9.8 metres per second squared over short distances, but objects which have a low density, such as feathers, have little or no acceleration.

4   For bodies moving with a constant acceleration, the average acceleration is the constant value of the acceleration, and since from para. 1,

$$a = \frac{v-u}{t}, \text{ then } a \times t = v - u \text{ or } \boldsymbol{v = u + at},$$

where
u is the initial velocity in m/s,
v is the final velocity in m/s,
a is the constant acceleration in m/s^2, and
t is the time in s.
When symbol 'a' has a negative value, it is called **deceleration** or

retardation. The equation, $v = u + at$ is called an **equation of motion**.

5   When an object is pushed or pulled, a force is applied to the object. This force is measured in **newtons**, (**N**). The effects of pushing or pulling an object are:

(i) to cause a change in the motion of the object, and

(ii) to cause a change in the shape of the object.

If a change occurs in the motion of the object, that is, its speed changes from u to v, then the object accelerates. Thus, it follows that acceleration results from a force being applied to an object. If a force is applied to an object and it does not move, then the object changes shape, that is, deformation of the object takes place. Usually the change in shape is so small that it cannot be detected by just watching the object. However, when very sensitive measuring instruments are used, very small changes in dimensions can be detected.

6   A force of attraction exists between all objects. The factors governing the size of this force are the masses of the objects and the distances between their centres,

$$\left(F \propto \frac{m_1 m_2}{d^2} \right).$$

Thus, if a person is taken as one object and the earth as a second object, a force of attraction exists between the person and the earth. This force is called the **gravitational force** and is the force which gives a person a certain weight when standing on the earth's surface. It is also this force which gives freely falling objects a constant acceleration in the absence of other forces.

7   To make a stationary object move or to change the direction in which the object is moving requires a force to be applied externally to the object. This concept is known as **Newton's first law of motion** and may be stated as:

'*an object remains in a state of rest, or continues in a state of uniform motion in a straight line, unless it is acted on by an externally applied force*'.

8   Since a force is necessary to produce a change of motion, an object must have some resistance to a change in its motion. The force necessary to give a stationary pram a given acceleration is far less than the force necessary to give a stationary car the same acceleration. The resistance to a change in motion is called the **inertia** of an object and the amount of inertia depends on the mass of the object. Since a car has a much larger mass than a pram, the inertia of a car is much larger than that of a pram.

9   **Newton's second law of motion** may be stated as:

'*the acceleration of an object acted upon by an external force is proportional to the force and is in the same direction as the force.*'

Thus, force ∝ acceleration
or force = a constant × acceleration, this constant of proportionality being the mass of the object, i.e.

force = mass × acceleration.

The unit of force is the newton (N) and is defined in terms of mass and acceleration. One newton is the force required to give a mass of 1 kilogram an acceleration of 1 metre per second squared. Thus:

$$F = ma \ , \text{ where}$$

F is the force in newtons (N), m is the mass in kilograms (kg) and a is the acceleration in metres per second squared (m/s^2), i.e.

$$1 \text{ N} = 1 \ \frac{\text{kg m}}{\text{s}^2}$$

10   **Newton's third law of motion** may be stated as:

'*for every force, there is an equal and opposite reacting force*'.

Thus, an object on, say, a table, exerts a downward force on the table and the table exerts an equal upward force on the object, known as a **reaction force** or just a **reaction**. When an object is accelerating, the force due to the inertia of the body ($F = ma$) is a reaction force acting in the opposite direction to the motion of the object.

11   When an object moves in a circular path at constant speed, its direction of motion is continually changing and hence its velocity (which depends on **both magnitude and direction**) is also continually changing. Since acceleration is the

$$\frac{\text{change in velocity}}{\text{change in time}} \ ,$$

the object has an acceleration. Let the object be moving with a constant angular velocity of ω and a tangential velocity of magnitude v and let the change of velocity for a small change of angle of θ ($= \omega t$) be V.

Then, $v_2 - v_1 = V$.

The vector diagram is shown in *Figure 26.2(b)* and since the magnitudes of v_1 and v_2 are the same, i.e. v, the vector diagram is also an isosceles triangle.

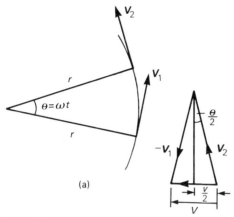

Figure 26.2

Bisecting the angle between v_2 and v_1 gives:

$$\sin \frac{\theta}{2} = \frac{V/2}{v_2} = \frac{V}{2v}$$

i.e. $V = 2v \sin \frac{\theta}{2}$ (1)

Since $\theta = \omega t$, $t = \dfrac{\theta}{\omega}$ (2)

Dividing (1) by (2) gives:

$$\frac{V}{t} = \frac{2v \sin \dfrac{\theta}{2}}{\dfrac{\theta}{\omega}} = \frac{v\omega \sin \dfrac{\theta}{2}}{\dfrac{\theta}{2}}$$

For small angles, $\dfrac{\sin \dfrac{\theta}{2}}{\dfrac{\theta}{2}}$ is very nearly equal to unity.

Hence, $\dfrac{V}{t} = \dfrac{\text{change of velocity}}{\text{change of time}} = \text{acceleration}$, $a = v\omega$

But, $\omega = \dfrac{v}{r}$, thus $v\omega = v \cdot \dfrac{v}{r} = \dfrac{v^2}{r}$

That is, the acceleration a is v^2/r and is towards the centre of the circle of motion (along V). It is called the **centripetal acceleration**. If the mass of the rotating object is m, then by Newton's second law the **centripetal force** is mv^2/r, and its direction is towards the centre of the circle of motion.

27 Linear momentum and impulse

1   (i) The **momentum** of a body is defined as the product of its mass and its velocity,

i.e. **momentum $= mu$**

where

$m =$ mass (in kg)

and

$u =$ velocity (in m/s).

    The unit of momentum is kg m/s.

(ii) Since velocity is a vector quantity, **momentum is a vector quantity**, i.e. it has both magnitude and direction.

2   (i) **Newton's first law of motion** states:

> '*a body continues in a state of rest or in a state of uniform motion in a straight line unless acted on by some external force.*'

Hence the momentum of a body remains the same provided no external forces act on it.

(ii) **The principle of conservation of momentum** for a closed system (i.e. one on which no external forces act) may be stated as:

> '*the total linear momentum of a system in any given direction is a constant*'.

(iii) The total momentum of a system before collision in a given direction is equal to the total momentum of the system after collision in the same direction. In *Figure 27.1*, masses m_1 and m_2 are

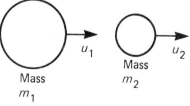

Figure 27.1

travelling in the same direction with velocity $u_1 > u_2$. A collision will occur, and applying the principle of conservation of momentum:

total momentum before impact = total momentum after impact

i.e. $m_1u_1 + m_2u_2 = m_1v_1 + m_2v_2$.

where v_1 and v_2 are the velocities of m_1 and m_2 after impact.

3   (i) **Newton's second law of motion** may be stated as:

    '*the rate of change of momentum is directly proportional to the applied force producing the change, and takes place in the direction of this force*'.

In the SI system, the units are such that:

    the applied force = rate of change of momentum

$$= \frac{\text{change of momentum}}{\text{change of time}} \qquad (1)$$

(ii) When a force is suddenly applied to a body due to either a collision with another body or being hit by an object such as a hammer, the time taken in equation (1) is very small and difficult to measure. In such cases, the total effect of the force is measured by the change of momentum it produces.

(iii) Forces which act for very short periods of time are called **impulsive forces**. The product of the impulsive force and the time during which it acts is called the **impulse** of the force and is equal to the change of momentum produced by the impulsive force,

    i.e. **impulse = applied force × time = change in linear momentum**.

(iv) Examples where impulsive forces occur, include when a gun recoils and when a free-falling mass hits the ground. Solving problems associated with such occurrences often requires the use of the equation of motion:

$$v^2 = u^2 + 2as.$$

4   When a pile is being hammered into the ground, the ground resists the movement of the pile and this resistance is called a **resistive force**.

Newton's third law of motion may be stated as:

    '*for every force there is an equal and opposite force*'.

The force applied to the pile is the resistive force. The pile exerts an equal and opposite force on the ground.

5   In practised when impulsive forces occur, momentum is not entirely conserved and some energy is changed into heat, noise, and so on.

28 Linear and angular motion

1 The unit of angular displacement is the **radian**, where one radian is the angle subtended at the centre of a circle by an arc equal in length to the radius, see *Figure 28.1*.

The relationship between angle in radians (θ), arc length (s) and radius of a circle (r) is:

$$s = r\theta \qquad (1)$$

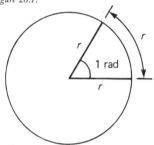

Figure 28.1

Since the arc length of a complete circle is $2\pi r$ and the angle subtended at the centre is $360°$, then, from equation (1), for a complete circle, $2\pi r = r\theta$ or $\theta = 2\pi$ radians.

Thus, 2π radians corresponds to $360°$ \qquad (2)

2 (i) **Linear velocity**, v, is defined as the rate of change of linear displacement, s, with respect to time, t, and for motion in a straight line:

linear velocity $= \dfrac{\text{change of distance}}{\text{change of time}}$,

i.e. $v = \dfrac{s}{t}$ \qquad (3)

The unit of linear velocity is metres per second (m/s).

(ii) **Angular velocity**

The speed of revolution of a wheel or a shaft is usually measured in revolutions per minute or revolutions per second but these units do not form part of a coherent system of units. The basis used in SI units is the angle turned through in one second.

Angular velocity is defined as the rate of change of angular displacement, θ, with respect to time, t, and for an object rotating about a fixed axis at a constant speed:

$$\text{angular velocity} = \frac{\text{angle turned through}}{\text{time taken}},$$

i.e. $\omega = \dfrac{\theta}{t}$ \hfill (4)

The unit of angular velocity is radians per second (rad/s). An object rotating at a constant speed of n revolutions per second subtends an angle of $2\pi n$ radians in one second, that is, its angular velocity

$$\boldsymbol{\omega = 2\pi n \ \textbf{rad/s}} \hfill (5)$$

(iii) From equation (1), $s = r\theta$ and from equation (4), $\theta = \omega t$, hence

$$s = r\omega t \text{ or } \frac{s}{t} = \omega r.$$

However, from equation (3),

$v = \dfrac{s}{t}$, hence $\boldsymbol{v = \omega r}$ \hfill (6)

Equation (6) gives the relationship between linear velocity, v, and angular velocity, ω.

3 (i) **Linear acceleration**, a, is defined as the rate of change of linear velocity with respect to time. For an object whose linear velocity is increasing uniformly:

$$\text{linear acceleration} = \frac{\text{change of linear velocity}}{\text{time taken}}$$

i.e. $a = \dfrac{v_2 - v_1}{t}$ \hfill (7)

The unit of linear acceleration is metres per second squared (m/s^2). Rewriting equation (7) with v_2 as the subject of the formula, gives:

$$v_2 = v_1 + at \hfill (8)$$

(ii) **Angular acceleration**, α, is defined as the rate of change of angular velocity with respect to time. For an object whose angular velocity is increasing uniformly:

$$\text{angular acceleration} = \frac{\text{change of angular velocity}}{\text{time taken}}$$

that is, $\alpha = \dfrac{\omega_2 - \omega_1}{t}$ \hfill (9)

The unit of angular acceleration is radians per second squared (rad/s^2). Rewriting equation (9) with ω_2 as the subject of the formula gives:

$$\omega_2 = \omega_1 + \alpha t \tag{10}$$

(iii) From equation (6), $v = \omega r$. For motion in a circle having a constant radius r, v_2 in equation (7) is given by $v_2 = \omega_2 r$ and $v_1 = \omega_1 r$, hence equation (7) can be written as:

$$a = \frac{\omega_2 r - \omega_1 r}{t} = \frac{r(\omega_2 - \omega_1)}{t}$$

But from equation (9), $\dfrac{\omega_2 - \omega_1}{t} = \alpha$

Hence $\boldsymbol{a = r\alpha}$ $\hspace{2cm}$ (11)

Equation (11) gives the relationship between linear acceleration a and angular acceleration α.

4 (i) From equation (3), $s = vt$, and if the linear velocity is changing uniformly from v_1 to v_2, then

$\quad s = $ mean linear velocity × time

i.e. $s = \left(\dfrac{v_1 + v_2}{2}\right)t$ $\hspace{2cm}$ (12)

(ii) From equation (4), $\theta = \omega t$, and if the angular velocity is changing uniformly from ω_1 to ω_2, then

$\quad \theta = $ mean angular velocity × time,

i.e. $\theta = \left(\dfrac{\omega_1 + \omega_2}{2}\right)t$ $\hspace{2cm}$ (13)

(iii) Two further equations of linear motion may be derived from equations (8) and (11):

$$s = v_1 t + \frac{1}{2} a t^2 \tag{14}$$

and $v_2^2 = \boldsymbol{v_1^2 + 2as}$ $\hspace{2cm}$ (15)

(iv) Two further equations of angular motion may be derived from equations (10) and (12):

$$\theta = \omega_1 t + \frac{1}{2} \alpha t^2 \tag{16}$$

and $\omega_2^2 = \omega_1^2 + 2\alpha\theta$ $\hspace{2cm}$ (17)

Table 28.1

s = arc length (m)	r = radius of circle (m)
t = time (s)	θ = angle (rad)
v = linear velocity (m/s)	ω = angular velocity (rad/s)
v_1 = initial linear velocity (m/s)	ω_1 = initial angular velocity (rad/s)
v_2 = final linear velocity (m/s)	ω_2 = final angular velocity (rad/s)
a = linear acceleration (m/s^2)	α = angular acceleration (rad/s^2)
n = speed of revolutions (revolutions per second)	

Equation No.	Linear motion	Angular motion
1	$s = r\theta$ m	
2		2π rad = 360°
3 and 4	$v = \dfrac{s}{t}$ m/s	$\omega = \dfrac{\theta}{t}$ rad/s
5		$\omega = 2\pi n$ rad/s
6	$v = \omega r$ m/s	
8 and 10	$v_2 = (v_1 + at)$ m/s	$\omega_2 = (\omega_1 + \alpha t)$ rad/s
11	$a = r\alpha$ m/s^2	
12 and 13	$s = \left(\dfrac{v_1 + v_2}{2}\right) t$ m	$\theta = \left(\dfrac{\omega_1 + \omega_2}{2}\right) t$ rad
14 and 16	$s = (v_1 t + \tfrac{1}{2} at^2)$ m	$\theta = (\omega_1 t + \tfrac{1}{2}\alpha t^2)$ rad
15 and 17	$v_2{}^2 = (v_1{}^2 + 2as)$ (m/s)^2	$\omega_2{}^2 = (\omega_1{}^2 + 2\alpha\theta)$ (rad/s)^2

5   *Table 28.1* summarises the principal equations of linear and angular motion for uniform changes in velocities and constant accelerations and also gives the relationship between linear and angular quantities.

6   A vector quantity is represented by a straight line lying along the line of action of the quantity and having a length which is proportional to the size of the quantity. Thus **ab** in *Figure 28.2* represents a velocity of 20 m/s, whose line of action is due west.
    The bold letters, **ab**, indicate a vector quantity and the order of the letters indicate that the time of action is from a to b.

7   Consider two aircraft A and B flying at a constant altitude, A

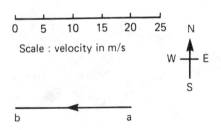

Figure 28.2

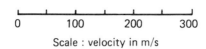

Scale : velocity in m/s

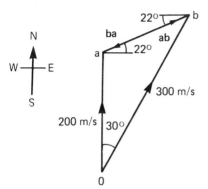

Figure 28.3

travelling due north at 200 m/s and B travelling 30° east of north, written N 30° E, at 300 m/s, as shown in *Figure 28.3*. Relative to a fixed point o, **oa** represents the velocity of A and **ob** the velocity of B. The velocity of B relative to A, that is the velocity at which B seems to be travelling to an observer on A, is given by **ab**, and by measurement is 160 m/s in a direction E 22° N.

The velocity of A relative to B, that is, the velocity at which A seems to be travelling to an observer on B, is given by **ba** and by measurement is 160 m/s in a direction W 22° S.

29 Friction

1   When an object, such as a block of wood, is placed on a floor and sufficient force is applied to the block, the force being parallel to the floor, the block slides across the floor. When the force is removed, motion of the block stops; thus there is a force which resists sliding. This force is called **dynamic** or **sliding friction**. A force may be applied to the block which is insufficient to move it. In this case, the force resisting motion is called the **static friction** or **striction**. Thus there are two categories into which a frictional force may be split:
(i) dynamic or sliding friction force which occurs when motion is taking place, and
(ii) static friction force which occurs before motion takes place.
2   There are three factors which affect the size and direction of frictional forces.
(i) The size of the frictional force depends on the type of surface (a block of wood slides more easily on a polished metal surface than on a rough concrete surface).
(ii) The size of the frictional force depends on the size of the force acting at right angles to the surfaces in contact, called the **normal force**. Thus, if the weight of a block of wood is doubled, the frictional force is doubled when it is sliding on the same surface.
(iii) The direction of the frictional force is always opposite to the direction of motion. Thus the frictional force opposes motion, as shown in *Figure 29.1*.
3   The **coefficient of friction**, μ, is a measure of the amount of friction existing between two surfaces. A low value of coefficient of friction indicates that the force required for sliding to occur is less

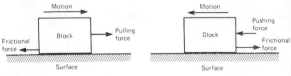

Figure 29.1

than the force required when the coefficient of friction is high. The value of the coefficient of friction is given by

$$\mu = \frac{\text{frictional force, } (F)}{\text{normal force, } (N)}$$

Transposing gives: frictional force $= \mu \times$ normal force,

$$\boxed{F = \mu N}$$

The direction of the forces given in this equation are as shown in *Figure 29.2*. The coefficient of friction is the ratio of a force to a force, and hence has no units. Typical values for the coefficient of

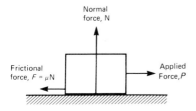

Figure 29.2

friction when sliding is occurring, i.e. the dynamic coefficient of friction are:

for polished oiled metal surfaces,	less than 0.1
for glass on glass,	0.4
for rubber on tarmac,	close to 1.0

4   In some applications, a low coefficient of friction is desirable, for example, in bearings, pistons moving within cylinders, on ski runs, and so on. However, for such applications as force being transmitted by belt drives and braking systems, a high value of coefficient is necessary.

Advantages and disadvantages of frictional forces

5   (a) Instances where frictional forces are an advantage include:

(i) Almost all fastening devices rely on frictional forces to keep them in place once secured, examples being screws, nails, nuts, clips and clamps.

(ii) Satisfactory operation of brakes and clutches rely on frictional forces being present.

(iii) In the absence of frictional forces, most accelerations along a horizontal surface are impossible. For example, a person's shoes just slip when walking is attempted and the types of a car just rotate with no forward motion of the car being experienced.

(b) **Disadvantages of frictional forces** include:

(i) Energy is wasted in the bearings associated with shafts, axles and gears due to heat being generated.

(ii) Wear is caused by friction, for example, in shoes, brake lining materials and bearings.

(iii) Energy is wasted when motion through air occurs (it is much easier to cycle with the wind rather than against it).

6 Two examples of **design implications** which arise due to frictional forces and how lubrication may or may not help are:

(i) Bearings are made of an alloy called white metal, which has a relatively low melting point. When the rotating shaft rubs on the white metal bearing, heat is generated by friction, often in one spot and the white metal may melt in this area, rendering the bearing useless. Adequate lubrication (oil or grease), separates the shaft from the white metal, keeps the coefficient of friction small and prevents damage to the bearing. For very large bearings, oil is pumped under pressure into the bearings and the oil is used to remove the heat generated, often passing through oil coolers before being recirculated. Designers should ensure that the heat generared by friction can be dissipated.

(ii) Wheels driving belts, to transmit force from one place to another, are used in many workshops. The coefficient of friction between the wheel and the belt must be high, and it may be increased by dressing the belt with a tar-like substance. Since frictional force is proportional to the normal force, a slipping belt is made more efficient by tightening it, thus increasing the normal and hence the frictional force. Designers should incorporate some belt tension mechanism into the design of such a system.

30 Waves

1   **Wave motion** is a travelling disturbance through a medium or through space, in which energy is transferred from one point to another without movement of matter.

2   **Examples where wave motion occurs** include:
  (i)   Water waves, such as are produced when a stone is thrown into a still pool of water;
 (ii)   waves on strings;
(iii)   waves on stretched springs;
 (iv)   sound waves;
  (v)   light waves (see page 225);
 (vi)   radio waves;
(vii)   infra-red waves, which are emitted by hot bodies;
(viii) ultra-violet waves, which are emitted by very hot bodies and some gas discharge lamps;
 (ix)   x-ray waves, which are emitted by metals when they are bombarded by high speed electrons;
  (x)   gamma-rays, which are emitted by radioactive elements.

Examples (i) to (iv) are **mechanical waves** and they require a medium (such as air or water) in order to move. Examples (v) to (x) are **electromagnetic waves** and do not require any medium — they can pass through a vacuum.

3   There are two types of wave, these being transverse and longitudinal waves.

(i) **Transverse waves** are where the particles of the medium move perpendicular to the direction of movement. For example, when a stone is thrown into a pool of still water, the ripple moves radially outwards but the movement of a floating object shows that the water at a particular point merely moves up and down. Light and radio waves are other examples of transverse waves.

(ii) **Longitudinal waves** are where the particles of the medium vibrate back and forth parallel to the direction of the wave travel. Examples include sound waves and waves in springs.

4   *Figure 30.1* shows a cross-section of a typical wave.

(i) **Wavelength** is the distance between two successive identical parts of a wave (for example, between two crests as shown in

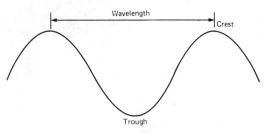

Figure 30.1

Figure 30.1). The symbol for wavelength is λ (Greek lambda) and its unit is metres.

(ii) **Frequency** is the number of complete waves (or cycles) passing a fixed point in one second. The symbol for frequency is f and its unit is the hertz, Hz.

(iii) The **velocity**, v of a wave is given by:

velocity = frequency × wavelength

$$\boxed{\text{i.e.} \quad v = f\lambda}$$

The unit of velocity is metres per second. Thus, for example, if BBC radio 4 is transmitted at a frequency of 200 kHz and a wavelength of 1500 m, the velocity of the radio wave v is given by

$$v = f\lambda = (200 \times 10^3)(1500) = 3 \times 10^8 \text{ m/s.}$$

5   **Reflection** is a change in direction of a wave while the wave remains in the same medium. There is no change in the speed of a reflected wave. All waves are reflected when they meet a surface through which they cannot pass. For example

(i)   light waves are reflected by mirrors;

(ii) water waves are reflected at the end of a bath or by a sea wall;

(iii) sound waves are reflected at a wall (which can produce an echo);

(iv) a wave reaching the end of a spring or string is reflected, and

(v)   television waves are reflected by satellites above the Earth.

    Experimentally, waves produced in an open tank of water may readily be observed to reflect off a sheet of glass placed at right angles to the surface of the water.

6   **Refraction** is a change in direction of a wave as it passes from one medium to another. All waves refract, and examples include:

222

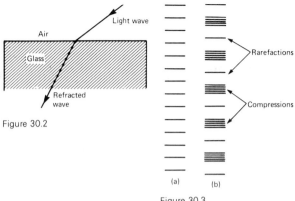

Figure 30.2

Figure 30.3

(i) a light wave changing its direction at the boundary between air and glass (see *Figure 30.2*),

(ii) sea waves refracting when reaching more shallow water, and

(iii) sound waves refracting when entering air of different temperature (see para. 8).

Experimentally, if one end of a water tank is made shallow the waves may be observed to travel more slowly in these regions and are seen to change direction as the wave strikes the boundary of the shallow area. The greater the change of velocity the greater is the bending or refraction.

7   A **sound wave** is a series of alternate layers of air, one layer at a pressure slightly higher than atmospheric, called compressions, and the other slightly lower, called refraction. In other words, **sound is a pressure wave**. *Figure 30.3(a)* represents layers of undisturbed air. *Figure 30.3(b)* shows what happens to the air when a sound wave passes.

Characteristics of sound waves

8   (i) Sound waves can travel through solids, liquids and gases, but not through a vacuum.

(ii) Sound has a finite (i.e. fixed) velocity, the value of which depends on the medium through which it is travelling. The velocity of sound is also affected by temperature. Some typical values for the velocity of sound are: air 331 m/s at 0°C, and 342 m/s at 18°C, water 1410 m/s at 20°C and iron 5100 m/s at 20°C.

(iii) Sound waves can be reflected, the most common example being an echo. Echo-sounding is used for charting the depth of the sea.

(iv) Sound waves can be refracted. This occurs, for example, when sound waves meet layers of air at different temperatures. If a sound wave enters a region of higher temperature the medium has different properties and the wave is bent as shown in *Figure 30.4*, which is typical of conditions that occur at night.

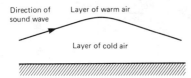

Figure 30.4

9   Sound waves are produced as a result of vibrations.

(i) In brass instruments, such as trumpets and trombones, or wind instruments, such as clarinets and oboes, sound is due to the vibration of columns of air.

(ii) In stringed instruments, such as guitars and violins, sound is produced by vibrating strings causing air to vibrate. Similarly, the vibration of vocal chords produces speech.

(iii) Sound is produced by a tuning fork due to the vibration of the metal prongs.

(iv) Sound is produced in a loudspeaker due to vibrations in the cone.

10   The pitch of a sound depends on the frequency of the vibration; the higher the frequency, the higher is the pitch. The frequency of sound depends on the form of the vibrator. The valves of a trumpet or the slide of a trombone lengthen or shorten the air column and the fingers alter the length of strings on a guitar or violin. The shorter the air column or vibrating string the higher the frequency and hence pitch. Similarly, a short tuning fork will produce a higher pitch note than a long tuning fork.

Frequencies between about 20 Hz and 20 kHz can be perceived by the human ear.

31 Light rays

1 (i) Light is an electromagnetic wave (see page 221) and the straight line paths followed by very narrow beams of light, along which light energy travels, are called **rays**.
(ii) The behaviour of light rays may be investigated by using a **ray-box**. This consists merely of a lamp in a box containing a narrow slit which emits rays of light.
(iii) Light always travels in straight lines although its direction can be changed by reflection or refraction.

Reflection of light

2 *Figure 31.1* shows a ray of light, called the incident ray, striking a plane mirror at O, and making an angle i with the normal, which is a line drawn at right angles to the mirror at O. i is called the **angle of incidence**. The light ray reflects as shown making an angle r with the normal. r is called the **angle of reflection**. There are **two laws of reflection**:
(i) The angle of incidence is equal to the angle of reflection (i.e. $i = r$ in *Figure 31.1*).
(ii) The incident ray, the normal at the point of incidence and the reflected ray all lie in the same plane.

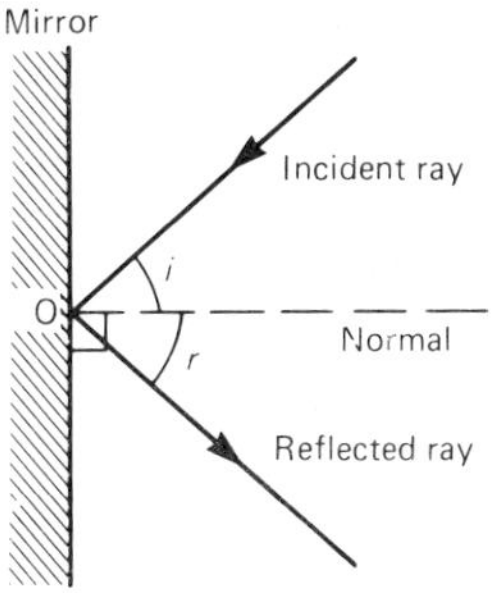

Figure 31.1

A **simple periscope arrangement** is shown in *Figure 31.2*. A ray of light from O strikes a plane mirror at an angle of 45° at point P. Since from the laws of reflection the angle of incidence i is equal to the angle of reflection r then $i = r = 45°$. Thus angle OPQ = 90° and the light is reflected through 90°. The ray then strikes another mirror at 45° at point Q. Thus a = b = 45°, angle PQR = 90° and the light ray is again reflected through 90°. Thus the light from O finally travels in the direction QR, which is parallel to OP, but displaced by the distance PQ. The arrangement thus acts as a periscope.

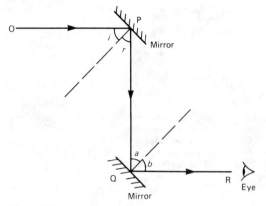

Figure 31.2

Refraction of light

3 (i) When a ray of light passes from one medium to another
the light undergoes a change in direction. This displacement of
light rays is called **refraction**.

(ii) *Figure 31.3* shows the path of a ray of light as it passes through
a parallel sided glass block. The incident ray AB which has an

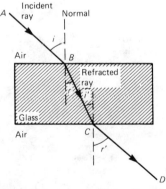

Figure 31.3

angle of incidence i enters the glass block at B. The direction of the ray changes to BC such that the angle r is less than angle i. r is called the angle of refraction. When the ray emerges from the glass at C the direction changes to CD, angle r' being greater than i. The final emerging ray CD is parallel to the incident ray AB.

(iii) In general, when entering a more dense medium from a less dense medium, light is refracted towards the normal and when it passes from a dense to a less dense medium it is refracted away from the normal.

4　(i) **Lenses** are pieces of glass or other transparent material with a spherical surface on one or both sides. When light is passed through a lens it is refracted.

(ii) Lenses are used in spectacles, magnifying glasses and microscopes, telescopes, cameras and projectors.

(iii) There are a number of different shaped lenses and two of the most common are shown in *Figure 31.4*. *Figure 31.4* shows a **bi-convex lens**, so called since both its surfaces curve outwards.

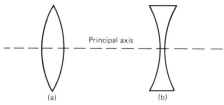

Figure 31.4

Figure 31.4(b) shows a **bi-concave lens**, so called since both of its surfaces curve inwards. The line passing through the centre of curvature of the lens surface is called the **principal axis**.

5　(i) *Figure 31.5* shows a number of parallel rays of light passing through a bi-convex lens. They are seen to converge at a point F on the principal axis.

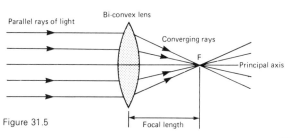

Figure 31.5

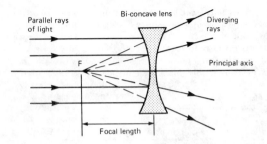

Figure 31.6

(ii) *Figure 31.6* shows parallel rays of light passing through a bi-concave lens. They are seen to diverge such that they appear to come from a point F which lies between the source of light and the lens, on the principal axis.

(iii) In both *Figure 31.5* and *Figure 31.6*, F is called the **principal focus** or the **focal point**, and the distance from F to the centre of the lens is called the **focal length** of the lens.

6   An **image** is the point from which reflected rays of light entering the eye appear to have originated. If the rays actually pass through the point then a **real image** is formed. Such images can be formed on a screen. *Figure 31.7* illustrates how the eye collects rays from an object after reflection from a plane mirror. To the eye, the rays appear to come from behind the mirror and the eye sees what seems to be an image of the object as far behind the mirror as the object is in front. Such an image is called a **virtual image** and this type cannot be shown on a screen.

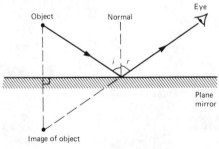

Figure 31.7

7   Lenses are important since they form images when an object emitting light is placed at an appropriate distance from the lens.

(a) **Bi-convex lenses**

(i) *Figure 31.8* shows an object O (a source of light) at a distance of more than twice the focal length from the

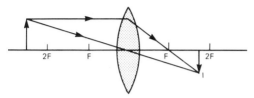

Figure 31.8

lens. To determine the position and size of the image two rays only are drawn, one parallel with the principal axis and the other passing through the centre of the lens. The image, I, produced is real, inverted (i.e. upside down), smaller than the object (i.e. diminished) and at a distance between one and two times the focal length from the lens. This arrangement is used in a **camera**.

(ii) *Figure 31.9* shows an object O at a distance of twice the focal length from the lens. This arrangement is used in a **photocopier**.

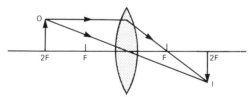

Figure 31.9

(iii) *Figure 31.10* shows an object O at a distance of between one and two focal lengths from the lens. The image I is real, inverted, magnified (i.e. greater than the object) and at a distance of more than twice the focal length from the lens. This arrangement is used in a **projector**.

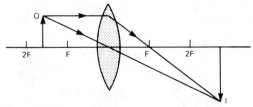

Figure 31.10

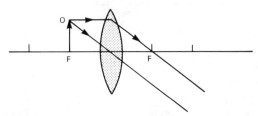

Figure 31.11

(iv) *Figure 31.11* shows an object O at the focal length of the lens. After passing through the lens the rays are parallel. Thus the image I can be considered as being found at infinity and being real, inverted and very much magnified. This arrangement is used in a **spotlight**.

(v) *Figure 31.12* shows an object O lying inside the focal length of the lens. The image I is virtual, since the rays

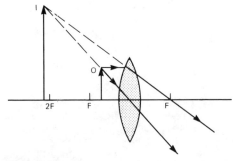

Figure 31.12

of light only appear to come from it, is on the same side of the lens as the object, is upright and magnified. This arrangement is used in a **magnifying glass**.

(b) Bi-concave lenses

For a bi-concave lens, as shown in *Figure 31.13*, the object O can be any distance from the lens and the image I

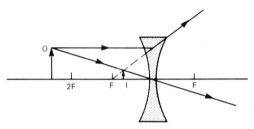

Figure 31.13

formed is virtual, upright, diminished and is found on the same side of the lens as the object. This arrangement is used in some types of **spectacles**.

8   A **compound microscope** is able to give large magnification by the use of two (or more) lenses. An object O is placed outside the focal length F_O of a bi-convex lens, called the objective lens (since it is near to the object), as shown in *Figure 31.14*. This produces a real, inverted, magnified image I_1. This

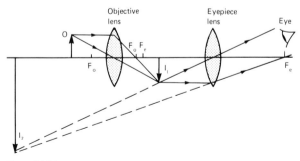

Figure 31.14

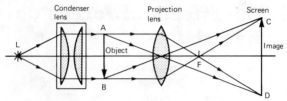

Figure 31.15

image then acts as the object for the eyepiece lens (i.e. the lens nearest the eye), and falls inside the focal length F_O of this lens. The eyepiece lens then produces a magnified, virtual, inverted image I_2 as shown in *Figure 31.14*.

9   A **simple projector arrangement** is shown in *Figure 31.15* and consists of a source of light and two lens systems. L is a brilliant source of light, such as a tungsten filament. One lens system called the condenser (usually consisting of two converting lenses as shown), is used to produce an intense illumination of the object AB, which is a slide transparency or film. The second lens, called the projection lens, is used to form a magnified, real, upright image of the illuminated object on a distant screen CD.

32 The effects of forces on materials

1   A **force** exerted on a body can cause a change in either the shape or the motion of the body. The unit of force is the **newton, N**.

2   No solid body is perfectly rigid and when forces are applied to it, changes in dimensions occur. Such changes are not always perceptible to the human eye since they are so small. For example, the span of a bridge will sag under the weight of a vehicle and a spanner will bend slightly when tightening a nut. It is important for engineers and designers to appreciate the effects of forces on materials, together with their mechanical properties.

3   The three main types of mechanical force that can act on a body are (i) tensile, (ii) compressive, and (iii) shear.

Tensile force

4   Tension is a force which tends to stretch a material, as shown in *Figure 32.1(a)*.

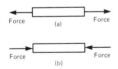

Figure 32.1

Examples include:

  (i)   the rope or cable of a crane carrying a load is in tension;
 (ii)   rubber bands, when stretched, are in tension;
(iii)   a bolt; when a nut is tightened, a bolt is under tension.

A tensile force, i.e. one producing tension, increases the length of the material on which it acts.

Compressive force

5   Compression is a force which tends to squeeze or crush a material, as shown in *Figure 32.1(b)*. Examples include:

(i) a pillar supporting a bridge is in compression;
(ii) the sole of a shoe is in compression;
(iii) the job of a crane is in compression.

A compressive force, i.e. one producing compression, will decrease the length of the material on which it acts.

Shear force

6 Shear is a force which tends to slide one face of the material over an adjacent face. Examples include:
(i) a rivet holding two plates together is in shear if a tensile force is applied between the plates, as shown in *Figure 32.2*.

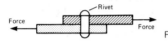

Figure 32.2

(ii) a guillotine cutting sheet metal, or garden shears each provide a shear force,
(iii) a horizontal beam is subject to shear force,
(iv) transmission joints on cars are subject to shear forces.

A shear force can cause a material to bend, slide or twist.

7 Forces acting on a material cause a change in dimensions and the material is said to be in a state of **stress**. Stress is the ratio of the applied force F to cross-sectional area A of the material. The symbol used for tensile and compressive stress is σ (Greek letter sigma). The unit of stress is the Pascal Pa, where 1 Pa=1 N/m^2.

Hence $$\boxed{\sigma = \frac{F}{A} \quad \textbf{Pa}}$$

where F is the force in newtons and A is the cross-sectional area in square metres.

For tensile and compressive forces, the cross-sectional area is that which is at right angles to the direction of the force.

8 The fractional change in a dimension of a material produced by a force is called the **strain**.

For a tensile or compressive force, strain is the ratio of the change of length to the original length. The symbol used for strain is ε (Greek epsilon). For a material of length l metres which changes in length by an amount x metres when subjected to stress,

$$\boxed{\varepsilon = \frac{x}{l}}$$

234

Strain is dimensionless and is often expressed as a percentage,

i.e. $$\boxed{\textbf{Percentage strain} = \frac{x}{l} \times 100\%}$$

9   (i) **Elasticity** is the ability of a material to return to its original shape and size on the removal of external forces.

(ii) **Plasticity** is the property of a material of being permanently deformed by a force without breaking. Thus if a material does not return to the original shape, it is said to be plastic.

(iii) Within certain load limits, mild steel, copper, polythene and rubber are examples of elastic materials; lead and plasticine are examples of plastic materials.

10   If a tensile force applied to a uniform bar of mild steel is gradually increased and the corresponding extension of the bar is measured, then provided the applied force is not too large, a graph depicting these results is likely to be as shown in *Figure 32.3*. Since the graph is a straight line,

extension is directly proportional to the applied force.

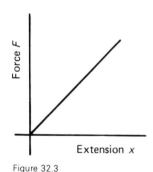

Figure 32.3

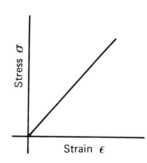

Figure 32.4

11   If the applied force is large, it is found that the material no longer returns to its original length when the force is removed. The material is then said to have passed its **elastic limit** and the resulting graph of force/extension is no longer a straight line (see para. 15).

Stress $\sigma = \dfrac{F}{A}$, from para. 7,

and since, for a particular bar, A can be considered as constant

then, $F \propto \sigma$.

Strain $\varepsilon = \dfrac{x}{l}$, from para. 8,

and since for a particular bar l is constant,

then, $x \propto \varepsilon$.

Hence for stress applied to a material below the elastic limit a graph of stress/strain will be as shown in *Figure 32.4*, and is a similar shape to the force/extension graph of *Figure 32.3*.

12 **Hooke's law** states:

'*Within the elastic limit, the extension of a material is proportional to the applied force.*'

It follows, from para. 11, that:

'*Within the elastic limit of a material, the strain produced is directly proportional to the stress producing it.*'

13 Within the elastic limit, stress $\propto$ strain

Hence, stress = (a constant) × strain

This constant of proportionality is called **Young's Modulus of Elasticity** and is given by the symbol E. The value of E may be determined from the gradient of the straight line portion of the stress/strain graph. The dimensions of E are pascals (the same as for stress, since strain is dimensionless).

$$E = \frac{\sigma}{\varepsilon} \ \textbf{Pa}$$

Some typical values for Young's modulus of elasticity, E, include: aluminium 70 GPa (i.e. 70×10^9 Pa), brass 100 GPa, copper 110 GPa, diamond 1200 GPa, mild steel 210 GPa, lead 18 GPa, tungsten 410 GPa, cast iron 110 GPa, zinc 110 GPa.

14 A material having a large value of Young's modulus is said to have a high value of stiffness, where stiffness is defined as:

$$\textbf{Stiffness} = \frac{\textbf{force } F}{\textbf{extension } x}$$

For example, mild steel is much stiffer than lead.

Since $E = \dfrac{\sigma}{\varepsilon}$ and $\sigma = \dfrac{F}{A}$ and $\varepsilon = \dfrac{x}{l}$

then $E = \dfrac{F/A}{x/l}$, i.e. $E = \dfrac{Fl}{Ax} = \left(\dfrac{F}{x}\right)\left(\dfrac{l}{A}\right)$

i.e. $E = (\textbf{stiffness}) \times \left(\dfrac{l}{A}\right)$

Stiffness $(=F/x)$ is also the gradient of the force/extension graph

hence: $E = (\textbf{gradient of force/extension graph})\left(\dfrac{l}{A}\right)$

Since l and A for a particular specimen are constant, the greater Young's modulus the greater the stiffness.

15   A **tensile test** is one in which a force is applied to a specimen of a material in increments and the corresponding extension of the specimen noted. The process may be continued until the specimen breaks into two parts and this is called testing to destruction. The testing is usually carried out using a universal testing machine which can apply either tensile or compressive forces to a specimen in small, accurately measured steps. BS 18 gives the standard procedure for such a test. Test specimens of a material are made to standard shapes and sizes and two typical test pieces are shown in *Figure 32.5*. The results of a tensile test may be plotted on a load/extension graph and a typical graph for a mild steel specimen is shown in *Figure 32.6*.

(i) Between A and B is the region in which Hooke's law applies and stress is directly proportional to strain. The gradient of AB is

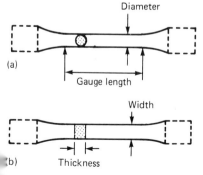

Figure 32.5

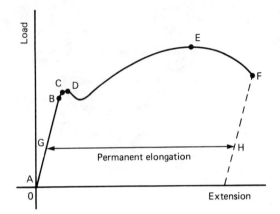

Figure 32.6

used when determining Young's modulus of elasticity (see para. 14).

(ii) Point B is the **limit of proportionality** and is the point at which stress is no longer proportional to strain when a further load is applied.

(iii) Point C is the **elastic limit** and a specimen loaded to this point will effectively return to its original length when the load is removed, i.e. there is negligible permanent extension.

(iv) Point D is called the **yield point** and at this point there is a sudden extension with no increase in load. The yield stress of the material is given by:

$$\text{yield stress} = \frac{\text{load where yield begins to take place}}{\text{original cross-sectional area}}$$

The yield stress gives an indication of the ductility of the material (see para. 16).

(v) Between points D and E extension takes place over the whole gauge length of the specimen.

(vi) Point E gives the maximum load which can be applied to the specimen and is used to determine the ultimate tensile strength (UTS) of the specimen (often just called the tensile strength).

$$\text{UTS} = \frac{\text{maximum load}}{\text{original cross-sectional area}}$$

(vi) Between points E and F the cross-sectional area of the

238

specimen decreases, usually about half way between the ends, and a **waist** or **neck** is formed before fracture.

The **percentage reduction in area**

$$= \frac{\textbf{original cross-sectional area} - \textbf{final cross-section area}}{\textbf{original cross-section area}} \times \textbf{100}\%$$

The percentage reduction in area provides information about the malleability of the material (see para. 16).

The value of stress at point F is greater than at point E since although the load on the specimen is decreasing as the extension increases, the cross-sectional area is also reducing.

(viii) At point F the specimen fractures.

(ix) Distance GH is called the **permanent elongation** and

Percentage elongation

$$= \frac{\textbf{increase in length during test to destruction}}{\textbf{original length}} \times \textbf{100}\%$$

16 (i) **Ductility** is the ability of a material to be plastically deformed by elongation, without fracture. This is a property which enables a material to be drawn out into wires. For ductile materials such as mild steel, copper and gold, large extensions can result before fracture occurs with increasing tensile force. Ductile materials usually have a percentage elongation value of about 15% or more.

(ii) **Brittleness** is the property of a material manifested by fracture without appreciable prior plastic deformation. Brittleness is a lack of ductility and brittle materials, such as cast iron, glass, concrete, brick and ceramics, have virtually no plastic stage, the elastic stage being followed by immediate fracture. Little or no 'waist' occurs before fracture in a brittle material undergoing a tensile test, and there is no noticeable yield point.

(iii) **Malleability** is the property of a material whereby it can be shaped when cold by hammering or rolling. A malleable material is capable of undergoing plastic deformation without fracture.

Shear stress and strain

17 For a shear force the shear stress is equal to force/area, where the area is that which is parallel to the direction of the force. The symbol for shear stress is the Greek letter tau, τ.

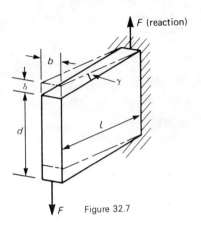

Figure 32.7

Hence from *Figure 32.7*, **shear stress**, $\tau = \dfrac{F}{bd}$

Shear strain is denoted by the Greek letter gamma, γ, and with reference to *Figure 32.7*,

shear strain $\gamma = \dfrac{\delta}{l}$.

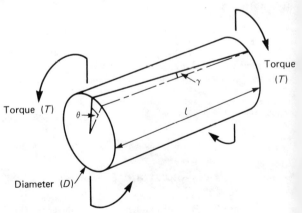

Figure 32.8

Modulus of rigidity, $G=\dfrac{\text{shear stress}}{\text{shear stress}}$,

i.e. $\boxed{G=\dfrac{\tau}{\gamma}.}$

For any metal the modulus of rigidity G is approximately 0.4 of the modulus of elasticity, E.

Torsional stress and strain

18   With reference to *Figure 32.8.*

$$\boxed{\dfrac{\tau}{\gamma}=\dfrac{T}{J}=\dfrac{G\theta}{l}}$$

where

$\tau=$ shear stress at radius r

$T=$ torque on shaft

$J=$ polar second moment of area of section of shaft

$G=$ modulus of rigidity

$\theta=$ angle of twist (radians) in a length l of shaft

(see *Figure 32.8*).

The polar second moment of area of a solid shaft is:

$$J=\dfrac{\pi D^4}{32}$$

where D is the diameter.

The polar second moment of area of a hollow shaft is:

$$J=\dfrac{\pi}{32}\,(D^4-d^4)$$

where $D=$ external diameter and $d=$ internal diameter.

33 Hardness and impact tests

Hardness tests

1   The **hardness** of a material may be defined in the following ways:
  (i)   the ability to scratch other materials;
 (ii)   the ability to resist scratching;
(iii)   the ability to resist plastic spreading under indentation;
 (iv)   the ability to resist elastic deformation under indentation;
  (v)   the ability to resist deformation by rolling.

2   Hardness tests are based on pressing a hard substance, such as a diamond or a steel sphere having known dimensions, into the material under test. The hardness can be determined from the size of indentation made for a known load. The three principal hardness tests are:
  (a)   the Brinell test,
  (b)   the Vickers test, and
  (c)   the Rockwell test.

3   In a standard **Brinell test**, a hardened steel ball having a diameter of 10 mm is squeezed into the material under a load of 3000 kg. The diameter of the indentation produced is measured under a microscope. The Brinell hardness number, H_B, is given by:

$$H_B = \frac{\text{load}}{\text{spherical area of indent}}$$

$$= \frac{F}{\frac{\pi D}{2}[D - \sqrt{(D^2 - d^2)}]}$$

where

  F is the load on kilogram (usually 3000 kg),
  D is the diameter of the steel ball in millimetres (usually 10 mm)
  d is the diameter of the indentation in millimetres.

Variations on the standard test include smaller loads used for soft materials, balls of different diameters (usually restricted to 1, 2 and

5 mm) and the steel ball being replaced by one made of tungsten-carbide for use with very hard materials. Values of Brinell hardness number vary from about 900 for very hard materials having an equivalent tensile strength for steel of 3000 MPa, down to about 100 for materials having an equivalent tensile strength for steel of about 350 MPa. The approximate relationship between tensile strength and hardness is discussed in para. 4.

The following precautions are required when carrying out a Brinell test.

(i) The material must be sufficiently wide and thick. The impression must have its centre not less than two and a half times its diameter from any edge. The thickness must be at least ten times the depth of the impression, this depth being given by

$$\frac{F}{\pi D H}$$

where H is the estimated Brinell hardness number.
(ii) The surface of the material should, if possible, be ground flat and polished.
(iii) The load should be held for 15 s
(iv) Two diameters of the impression at right angles should be read and their mean used in the calculation.
(v) When stating the result of the test the ball number and load should be stated, e.g. H 10/3000 = 410.

Special machines are made but a Brinell test can be carried out on most universal testing machines.

4   For materials of the same quality and for families of materials an approximate direct proportional relationship seems to exist between tensile strength and Brinell hardness number. For example, a nickel-chrome steel which is hardened and then tempered to various temperatures has tensile strengths varying from 1900 MPa to 1070 MPa as the Brinell hardness number varies from 530 to 300. A constant of proportionality k for: tensile strength $= (k \times \text{hardness})$ in this case is 3.57 for all tempering temperatures. Similarly, for a family of carbon steels, the tensile strength varies from 380 MPa to 790 MPa as the carbon content increases. The Brinell hardness number varies from 115 to 230 over the same range of carbon values and the constant of proportionality in this case is 3.35. Because of the general approximate relationship between tensile strength and hardness, tables exist relating these quantities, the tables usually being based on a constant of proportionality of about 3.35.

5   In a **Vickers diamond pyramid hardness test**, a square-

based diamond pyramid is pressed into the material under test. The angle between opposite faces of the diamond is 136° and the load applied is one of the values 5, 10, 30, 50 or 120 kg, depending on the hardness of the material. The Vickers diamond hardness number, H_V, is given by:

$$H_V = \frac{\text{load}}{\text{surface area of indentation}} = \frac{F}{d^2/1.854}$$

where F is the load in kilograms and d is the length of the diagonal of the square of indentation in millimetres.

6   The **Rockwell hardness test** is mainly used for rapid routine testing of finished material, the hardness number being indicated directly on a dial. The value of hardness is based directly on the depth of indentation of either a steel ball or a cone shaped diamond with a spherically rounded tip, called a 'brale'. Whether the steel ball or brale is selected for use depends on the hardness of the material under test, the steel ball being used for materials having a hardness up to that of medium carbon steels.

Several different scales are shown on the dial, and can include Rockwell A to H scales together with Rockwell K, N and T scales. Examples of the scale used are:

Scale A: using a brale and a 60 kg load;
Scale B: using a brale and a 150 kg load;
Scale C: using a 1/16th inch steel ball and 100 kg load,
and so on.

The big **advantage** of the Rockwell test over Brinell and Vickers tests is the speed with which it can be made. As it is also independent of surface condition it is well suited to production line testing. British Standards, however, require hardness numbers to be based on the surface area of any indentation.

Other hardness tests

7   The Brinell, Vickers and Rockwell tests are examples of static hardness tests. Another example is the **Firth Hardometer test**, which is very similar to the Vickers test.

Examples of dynamic hardness tests are those using the **Herbert Pendulum Hardness Tester** and the **Shore Scleroscope**. The former uses an arched rocker resting on a steel or diamond pivot; hardness can be indicated by the time taken for ten single swings or by the difference between an initial angular displacement and the first swing. The Scleroscope is a portable apparatus in which a diamond-tipped hammer falls on to the material under test. The height of the rebound gives the hardness number.

Other non-destructive tests

8   Flaws inside a casting can be revealed by X-ray methods. Surface flaws can be revealed by electro-magnetic methods and by those using ultra-violet light. The former are applicable only to ferrous metals but the latter can be used for other metals and for other materials such as plastics and ceramics.

Impact tests

9   To give an indication of the toughness of a material, that is, the energy needed to fracture it, impact tests are carried out. Two such tests are the Izod test, principally used in Great Britain and the Charpy test which is widely used on the continent of Europe.

10   In an **Izod test**, a square test piece of side 10 mm and having a vee-notch of angle 45° machined along one side, is clamped firmly in a vice in the base of the Izod test machine. A heavy pendulum swings down to strike the specimen and fractures it. The difference between the release angle of the pendulum measured to the vertical and the overswing angle after fracturing the specimen is proportional to the energy expended in fracturing the specimen, and can be read from a scale on the testing machine. An Izod test is basically an acceptance test, that is, the value of impact energy absorbed is either acceptable or is not acceptable. The results of an Izod test cannot be used to determine impact strength under other conditions.

11   A **Charpy test** is similar to an Izod test, the only major difference being the method of mounting the test specimen and a capability of varying the mass of the pendulum. In the Izod test, the specimen is gripped at one end and is supported as a cantilever, compared with the specimen being supported at each end as a beam in the Charpy test. One other difference is that the notch is at the centre of the supported beam and faces away from the striker.

34 Centre of gravity and equilibrium

1 The **centre of gravity** of an object is a point where the resultant gravitational force acting on the body may be taken to act. For objects of uniform thickness lying in a horizontal plane, the centre of gravity is vertically in line with the point of balance of the object. For a thin uniform rod the point of balance and hence the centre of gravity is halfway along the rod, as shown in *Figure 34.1(a)*.

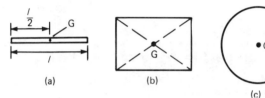

Figure 34.1

A thin flat sheet of a material of uniform thickness is called a **lamina** and the centre of gravity of a rectangular lamina lies at the point of intersection of its diagonals, as shown in *Figure 34.1(b)*. The centre of gravity of a circular lamina is at the centre of the circle, as shown in *Figure 34.1(c)*.

2 An object is in **equilibrium** when the forces acting on the object are such that there is no tendency for the object to move. The state of equilibrium of an object can be divided into three groups.

(a) If an object is in **stable equilibrium** and it is slightly disturbed by pushing or pulling (i.e. a disturbing force is applied), the centre of gravity is raised and when the distributing force is applied, the centre of gravity is original position. Thus a ball bearing in a hemispherical cup is in stable equilibrium, as shown in *Figure 34.2(a)*.

(b) An object is in **unstable equilibrium** if, when a disturbing force is applied, the centre of gravity is

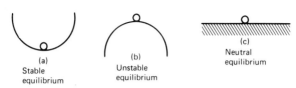

(a)
Stable
equilibrium

(b)
Unstable
equilibrium

(c)
Neutral
equilibrium

Figure 34.2

lowered and the object moves away from its original position. Thus, a ball bearing balanced on top of a hemispherical cup is in unstable equilibrium, as shown in *Figure 34.2(b)*.

(c) When an object in **neutral equilibrium** has a disturbing force applied, the centre of gravity remains at the same height and the object does not move when the disturbing force is removed. Thus, a ball bearing on a flat horizontal surface is in neutral equilibrium, as shown in *Figure 34.2(c)*.

35 Coplanar forces acting at a point

1   When forces are all acting in the same plane, they are called **coplanar**. When forces act at the same time and at the same point, they are called **concurrent/forces**.

2   Force is a vector quantity and thus has both a magnitude and a direction. A vector can be represented graphically by a line drawn to scale in the direction of the line of action of the force. Vector quantities may be shown by using bold, lower case letters, thus **ab** in *Figure 35.1* represents a force of 5 newtons acting in a direction due east.

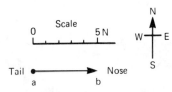

Figure 35.1

The resultant of two coplanar forces

3   For two forces acting at a point, there are three possibilities.
   (a) For forces acting in the same direction and having the same line of action, the single force having the same effect as both of the forces, called the **resultant force** or just the **resultant**, is the arithmetic sum of the separate forces. Forces of F_1 and F_2 acting at point P, as shown in *Figure 35.2(a)* have exactly the same effect on point P as force F shown in *Figure 35.2(b)*, where $F = F_1 + F_2$ and acts in the same direction as F_1 and F_2. Thus, F is the resultant of F_1 and F_2.
   (b) For forces acting in opposite directions along the same line of action, the resultant force is the arithmetic difference between the two forces. Forces of F_1 and F_2 acting at point P as shown in *Figure 35.3(a)*, have exactly

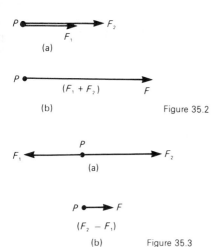

(a)

(b)

$(F_1 + F_2)$ F

Figure 35.2

(a)

$P \bullet\!\!\longrightarrow F$

$(F_2 - F_1)$

(b) Figure 35.3

the same effect on point P as force F shown in *Figure 35.3(b)*, where $F = F_2 - F_1$ and acts in the direction of F_2, since F_2 is greater than F_1. Thus F is the resultant of F_1 and F_2.

(c) When two forces do not have the same line of action, the magnitude and direction of the resultant force may be found by a procedure called vector addition of forces. There are two graphical methods of performing **vector addition**, known as the triangle of forces method and the parallelogram of forces method.

The triangle of forces method

4 (i) Draw a vector representing one of the forces, using an appropriate scale and in the direction of its line of action.
(ii) From the **nose** of this vector and using the same scale, draw a vector representing the second force in the direction of its line of action.
(iii) The resultant vector is represented in both magnitude and direction by the vector drawn from the **tail** of the first vector to the nose of the second vector.

Thus, for example, to determine the magnitude and direction of the resultant of a force of 15 N acting horizontally to the right and a force of 20 N, inclined at an angle of 60° to the 15 N force, using

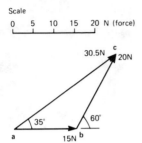

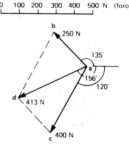

Figure 35.4

Figure 35.5

the triangle of forces method: With reference to *Figure 35.4* and using the above procedure:

(i) **ab** is drawn 15 units long horizontally;

(ii) from b, **bc** is drawn 20 units long, inclined at an angle of 60° to ab. (Note, in angular measure, an angle of 60° from ab means 60° in an anticlockwise direction.)

(iii) By measurement, the resultant **ac** is 30.5 units long inclined at an angle of 35° to ab.

Hence the resultant force is **30.5 N** inclined at an angle of **35°** to the 15 N force.

The parallelogram of forces method:

5   (i) Draw a vector representing one of the forces, using an appropriate scale and in the direction of its line of action.

(ii) From the **tail** of this vector and using the same scale draw a vector representing the second force in the direction of its line of action.

(iii) Complete the parallelogram using the two vectors drawn in (i) and (ii) as two sides of the parallelogram.

(iv) The resultant force is represented in both magnitude and direction by the vector corresponding to the diagonal of the parallelogram drawn from the tail of the vectors in (i) and (ii).

Thus, for example, to determine the magnitude and direction of the resultant of a 250 N force acting at an angle of 135° and a force of 400 N acting at an angle of −120°, using the parallelogram of force method: With reference to *Figure 35.5* and using the above procedure:

(i) **ab** is drawn at an angle of 135° and 250 units in length;

(ii) **ac** is drawn at an angle of $-120°$ and 400 units in length;

(iii) bc and cd are drawn to complete the parallelogram;

(iv) **ad** is drawn. By measurement **ad** is 413 units long at an angle of $-156°$.

Hence the resultant force is **413 N** at an angle of $-\mathbf{156°}$.

6   An alternative to the graphical methods of determining the resultant of two coplanar forces is by **calculation**. This can be achieved by trigonometry using the cosine rule and the sine rule, or by resolution of forces (see para. 9).

The resultant of more than two coplanar forces

7   For the three coplanar forces F_1, F_2 and F_3 acting at a point as shown in *Figure 35.6*, the vector diagram is drawn using the nose to tail method. The procedure is:

(i) Draw **oa** to scale to represent force F_1 in both magnitude and direction (see *Figure 35.7*).

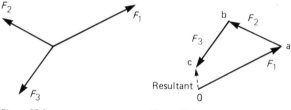

Figure 35.6   Figure 35.7

(ii) From the nose of **oa**, draw **ab** to represent force F_2.

(iii) From the nose of **ab**, draw **bc** to represent force F_3.

(iv) The resultant vector is given by length **oc** in *Figure 35.7*. The direction of resultant **oc** is from where we started, i.e. point o, to where we finished, i.e. point c. When acting by itself, the resultant force, given by **oc**, has the same effect on the point as forces F_1, F_2 and F_3 have when acting together. The resulting vector diagram of *Figure 35.7* is called the **polygon of forces**.

8   When three or more coplanar forces are acting at a point and the vector diagram closes, there is no resultant. The forces acting at the point are in **equilibrium**.

Resolution of forces

9   A vector quantity may be expressed in terms of its **horizontal and vertical components**. For example, a vector representing a force of 10 N at an angle of 60° to the horizontal is shown in *Figure 35.8*. If the horizontal line oa and the vertical line ab are constructed as shown, then oa is called the horizontal component of the 10 N force and ab the vertical component of the 10 N force.

From trigonometry,

$$\cos 60° = \frac{\text{oa}}{\text{ob}}.$$

Hence the horizontal

    component, oa = 10 cos 60°.

$$\sin 60° = \frac{\text{ab}}{\text{ob}}.$$

Hence the vertical

    component, ab = 10 sin 60°.

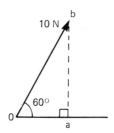

Figure 35.8

    This process is called '**finding the horizontal and vertical components of a vector**' or '**the resolution of a vector**', and can be used as an alternative to graphical methods for calculating the resultant of two or more coplanar forces acting at a point.

    For example, to calculate the resultant of a 10 N force acting at 60° to the horizontal and a 20 N force acting at − 30° to the horizontal (see *Figure 35.9*) the procedure is as follows:

(i) Determine the horizontal and vertical components of the 10 N force, i.e.

       horizontal component, oa = 10 cos 60° = 5.0 N
         vertical component, ab = 10 sin 60° = 8.66 N

(ii) Determine the horizontal and vertical components of the 20 N force, i.e.

       horizontal component, od = 20 cos (− 30°) = 17.32 N
         vertical component, dc = 20 sin (− 30°) = − 10.0 N.

(iii) Determine the total horizontal component, i.e.

    oa + od = 5.0 + 17.32 = 22.32 N

(iv) Determine the total vertical component, i.e.

    ab + cd = 8.66 + (− 10.0) = − 1.34 N

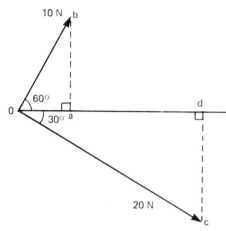

Figure 35.9

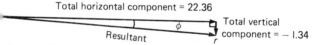

Figure 35.10

(v) Sketch the total horizontal and vertical components as shown in *Figure 35.10*. The resultant of the two components is given by length **or** and, by Pythagoras' theorem,

$$\text{or} = \sqrt{[(22.32)^2 + (1.34)^2]} = 22.36 \text{ N}$$

and using trigonometry,

$$\text{angle } \phi = \arctan\left[\frac{1.34}{22.32}\right] = 3° \ 26.$$

Hence the resultant of the 10 N and 20 N forces shown in *Figure 35.7* is 22.36 N at an angle of $-3°$ 26 to the horizontal.

The above example demonstrates the use of resolution of forces for calculating the resultant of two coplanar forces acting at a point. However, the method may be used for more than two forces acting at a point.

Summary

10   (a) To determine the resultant of two coplanar forces acting at a point, there are four methods commonly used. These are, by drawing:

      1 triangle of forces method,
      2 parallelogram of forces method,

and by calculation:

      3 use of cosine and sine rules,
      4 resolution of forces.

  (b) To determine the resultant of more than two coplanar forces acting at a point, there are two methods commonly used. These are, by drawing:

      1 polygon of forces method,

and by calculation:

      2 resolution of forces.

36 **Simply supported beams**

1   When using a spanner to tighten a nut, a force tends to turn
the nut in a clockwise direction. This turning effect of a force is
called the **moment of a force** or more briefly, just a **moment**.
The size of the moment acting on the nut depends on two factors:

  (a) the size of the force acting at right angles to the shank of
      the spanner, and

  (b) the perpendicular distance between the point of
      application of the force and the centre of the nut.

In general, with reference to *Figure 36.1*, the moment M of a force
acting at a point P = force × perpendicular distance between the
line of action of the force and P.

i.e. $\boxed{M = F \times d}$

The unit of a moment is the newton metre (Nm). Thus, if force F
in *Figure 36.1* is 7 N and distance d is 3 m, then the moment at P is
$7(\text{N}) \times 3(\text{m})$, i.e. 21 Nm.

2   If more than one force is acting on an object and the forces
do not act at a point, then the turning effect of the forces, that is,
the moment of the forces, must be considered.

Figure 36.2 shows a beam with its support (known as its pivot
or fulcrum), at P, acting vertically upwards, and forces F_1 and F_2
acting vertically downwards at distances a and b respectively from
the fulcrum.

A beam is said to be in **equilibrium** when there is no
tendency for it to move.

Figure 36.1

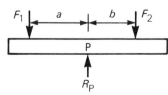

Figure 36.2

There are two conditions for equilibrium:
(i) the sum of the forces acting vertically downwards must be equal to the sum of the forces acting vertically upwards, i.e. for *Figure 36.2*,

$$R_p = F_1 + F_2, \text{ and}$$

(ii) the total moment of the forces acting on a beam must be zero; for the total moment to be zero:

'the sum of the clockwise moments about any point must be equal to the sum of the anticlockwise moments about that point'.

This statement is known as the **principle of moments**. Hence, taking moments about P in *Figure 36.2*, $F_2 \times b =$ the clockwise moment, and $F_1 \times a =$ the anticlockwise moment. Thus for equilibrium:

$$F_1 a = F_2 b.$$

3   (i) A **simply supported beam** is one which rests on two supports and is free to move horizontally.
(ii) Two typical simply supported beams having loads acting at given points on the beam (called **point loading**), as shown in *Figure 36.3*.

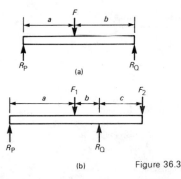

Figure 36.3

A man whose mass exerts a force F vertically downwards, standing on a wooden plank which is simply supported at its ends, may, for example, be represented by the beam diagram of *Figure 36.3(a)* if the mass of the plank is neglected. The forces exerted by the supports on the plank, R_P and R_Q, act vertically upwards, and are called **reactions**.
(iii) When the forces acting are all in one plane, the algebraic sum

of the moments can be taken about **any** point. For the beam in *Figure 36.3(a)*, at equilibrium:

(i) $R_P + R_Q = F$, and

(ii) taking moments about R_P, $Fa = R_Q b$

(Alternatively, taking moments about F, $R_P a = R_Q b$).

For the beam in *Figure 36.3(b)*, at equilibrium:

(i) $R_P + R_Q = F_1 + F_2$, and

(ii) taking moments about R_Q, $R_P(a + b) + F_2 c = F_1 b$.

(iv) Typical practical applications of simply supported beams with point loadings include bridges, beams in buildings and beds of machine tools.

For example, for the beam shown in *Figure 36.4*, the force acting on support A, R_A, and distance d are calculated as follows:

(Forces acting in an upward direction)
= (Forces acting in a downward direction)

i.e. $R_A + 40 = 10 + 15 + 30$

and $R_A = 10 + 15 + 30 - 40 = \mathbf{15\ N}$

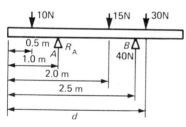

Figure 36.4

Taking moments about the left hand end of the beam and applying the principle of moments gives:

clockwise moments = anticlockwise moments

i.e. $(10 \times 0.5) + (15 \times 2.0) + (30d) = (15 \times 1.0) + (40 \times 2.5)$

i.e. $35 + 30d = 15$

from which, distance $d = \dfrac{115 - 35}{30} = \mathbf{2\dfrac{2}{3}\ m}$

Shearing force and bending moments

4   As stated in para. 3, for equilibrium of a beam, the forces to the left of any section such as X in *Figure 36.5*, must balance the forces to the right. Also the moment about X of the forces to the left must balance the moment about X of the forces to the right.

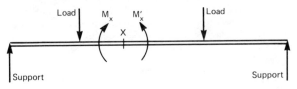

Figure 36.5

Although for equilibrium the forces and moments cancel, the magnitude and nature of these forces and moments are important as they determine both the stresses at X and the beam curvature and deflection. The resultant force to the left of X and the resultant force to the right of X (forces or components of forces transverse to the beam), constitute a pair of forces tending to shear the beam at this section. **Shearing force** is defined as the force transverse to the beam at a given section tending to cause it to shear at that section.

By convention, if the tendency is to shear as shown in *Figure 36.6(a)*, the shearing force is regarded as positive, i.e. $+F$; if the tendency to shear is as shown in *Figure 36.6(b)*, it is regarded as negative, i.e. $-F$.

5   The **bending moment** at a given section of a beam is defined as the resultant moment about that section of either all of

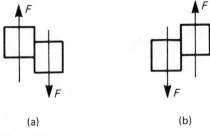

(a)   (b)

Figure 36.6

the forces to its left — or of all of the forces to its right. In *Figure 36.5* it is M_X or M'_X. These moments, clockwise to the left and anticlockwise to the right, will cause the beam to bend concave upwards, called '**sagging**'. By convention this is regarded as positive bending (i.e. the bending moment is a positive bending moment). Where the curvature produced is concave downwards (called '**hogging**'), the bending moment is regarded as negative.

The values of shearing force and bending moment will usually vary along a beam. Diagrams showing the shearing force and bending moment for all sections of a beam are called shearing force and bending moment diagrams respectively.

Shearing forces and shearing force diagrams are less important than bending moments, but can be very useful in giving pointers to the more important bending moment diagrams. For example, wherever the shearing force is zero, the bending moment will be a maximum or a minimum. The shearing force and bending moment diagrams for the beam shown in *Figure 36.7* are obtained as follows:

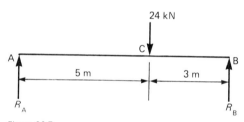

Figure 36.7

It is first necessary to calculate the reactions at A and B. The beam is simply-supported at A and B which means that it rests on supports at these points giving vertical reactions. The general conditions for equilibrium require that the resultant moment about any point must be zero, and total upward force must equal total downward force. Therefore, taking moments about A, the moment of R_B must balance the moment of the load at C:

$$R_B \times 8 \text{ m} = 24 \text{ kN} \times 5 \text{ m} = 120 \text{ kNm}$$

$$R_B = \frac{120 \text{ kNm}}{8 \text{ m}} = 15 \text{ kN, and}$$

$$R_A = 24 \text{ kN} - 15 \text{ kN} = 9 \text{ kN.}$$

Immediately to the right of A the shearing force is due to R_A and is therefore 9 kN. As this force to the left of the section considered, is upwards, the shearing force is positive. The shearing force is the same for all points between A and C as no other forces come on the beam between these points.

When a point to the right of C is considered, the load at C as well as R_A must be considered, or alternatively, R_B on its own. The shearing force is 15 kN, either obtained from $R_B = 15$ kN, or from load at $C - R_A = 15$ kN. For any point between C and B the force to the right is upwards and the shearing force is therefore negative. It should be noted that the shearing force changes suddenly at C.

The bending moment at A is zero, as there are no forces to its left. At a point 1 m to the right of A the moment of the only force R_A to the left of the point is $R_A \times 1$ m $= 9$ kNm. As this moment to the left is clockwise the bending moment is positive, i.e. it is $+9$ kNm. At points 2 m, 3 m, 4 m and 5 m to the right of A the bending moments are respectively:

$R_A \times 2$ m $= 9$ kN $\times 2$ m $= 18$ kNm
$R_A \times 3$ m $= 9$ kN $\times 3$ m $= 27$ kNm
$R_A \times 4$ m $= 9$ kN $\times 4$ m $= 36$ kNm
$R_A \times 5$ m $= 9$ kN $\times 5$ m $= 45$ kNm

All are positive bending moments.

For points to the right of C, the load at C as well as R_A must be considered or, more simply, R_B alone can be used. At points 5 m, 6 m and 7 m from A the bending moments are respectively:

$R_B \times 3$ m $= 15$ kN $\times 3$ m $= 45$ kNm
$R_B \times 2$ m $= 15$ kN $\times 2$ m $= 30$ kNm
$R_B \times 1$ m $= 15$ kN $\times 1$ m $= 15$ kNm

As these moments to the right of the points considered are anticlockwise they are all positive bending moments. At B the bending moment is zero as there is no force to its right. The results are summarised in the table below.

Distance from A (m)	0	1	2	3	4	5	6	7	8
Shearing force (kN)	+9	+9	+9	+9	+9	+9 / −15	−15	−15	−15
Bending moment (kNm)	0	+9	+18	+27	+36	+45	+30	+15	0

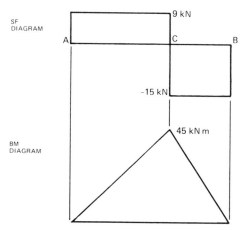

SF DIAGRAM

9 kN

A   C   B

-15 kN

BM DIAGRAM

45 kN m

Figure 36.8

Making use of the above values, the diagrams are as shown (in the usual manner) in *Figure 36.8*. A stepped shearing force diagram, with horizontal and vertical lines only, is always obtained when the beam carries concentrated loads only. A sudden change in shearing force occurs where the concentrated loads, including the reactions at supports, occur. For this type of simple loading the bending moment diagram always consists of straight lines, usually sloping. Sudden changes of bending moment cannot occur except in the unusual circumstances of a moment being applied to a beam as distinct from a load.

6   Bending stress

$$\frac{\sigma}{y} = \frac{M}{I} \left(= \frac{E}{R} \right)$$

where

σ = stress due to bending at distance y from the neutral axis;

M = bending moment;

I = second moment of area of section of beam about its neutral axis;

E = modulus of elasticity;

R = radius of curvature.

Section modulus $Z = I/y_{max}$.

The second moments of area of the beam sections most commonly met with are (about the central axis XX):

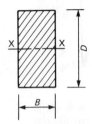

Figure 36.9

(a) **Solid rectangle** (*Figure 36.9*).

$$I = \frac{BD^3}{12}$$

(b) **Symmetrical hollow rectangle or *I* section** (*Figure 36.10*)

$$I = \frac{BD^3 - bd^3}{12}$$

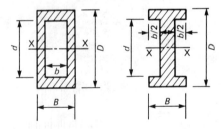

Figure 36.10

(c) **Solid rod** (*Figure 36.11*)

$$I = \frac{\pi D^4}{64}$$

(d) **Tube** (*Figure 36.12*)

$$I = \frac{\pi(D^4 - d^4)}{64}$$

The neutral axis of any section, where bending produces no strain and therefore no stress, always passes through the centroid of the

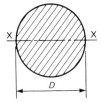

Figure 36.11

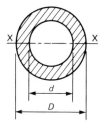

Figure 36.12

section. For the symmetrical sections listed above this means that for vertical loading the neutral axis is the horizontal axis of symmetry.

For example, let the maximum bending moment on a beam be 120 Nm. If the beam section is rectangular 18 mm wide and 36 mm deep, the maximum bending stress is calculated as follows:

Second moment of area of·section about the neutral axis,

$$I = \frac{bd^3}{12} = \frac{(18)(36)^3}{12} = 6.9984 \times 10^4 \text{ mm}^4$$

Maximum distance from neutral axis,

$$y = \frac{36}{2} = 18 \text{ mm}.$$

Since $\sigma/y = M/I$ then the maximum bending stress σ will occur where M and y have their maximum values, i.e.

$$\sigma = \frac{My}{I} = \frac{120 \text{ Nm} \times 18 \text{ mm}}{6.9984 \times 10^4 \text{ mm}^4}$$

$$= \frac{120 \text{ Nm} \times 18 \times 10^{-3} \text{ m}}{6.9984 \times 10^4 \times 10^{-12} \text{ m}^4}$$

$$= 30.86 \text{ MN/m}^2$$

$$= \textbf{30.86 MPa}.$$

37 Work, energy and power

1   Fuel, such as oil, coal, gas or petrol, when burnt, produces heat. Heat is a form of energy and may be used, for example, to boil water or to raise steam. Thus fuel is useful since it is a convenient method of storing energy, that is, **fuel is a source of energy**.

2   (i) If a body moves as a result of a force being applied to it, the force is said to do work on the body. The amount of work done is the product of the applied force and the distance, i.e.

Work done = force × distance moved in the direction of the force

(ii) The unit of work is the **joule, J**, which is defined as the amount of work done when a force of 1 Newton acts for a distance of 1 metre in the direction of the force.

Thus, 1 J = 1 Nm.

3   If a graph is plotted of experimental values of force (on the vertical axis) against distance moved (on the horizontal axis) a force-distance graph or work diagram is produced. **The area under the graph represents the work done.**

For example, a constant force of 20 N used to raise a load a height of 8 m may be represented on a force-distance graph as shown in *Figure 37.1(a)*. The area under the graph shown shaded, represents the work done.

Hence, work done = 20 N × 8 m = **160 J**

Similarly, a spring extended by 20 mm by a force of 500 N may be represented by the work diagram shown in *Figure 36.1(b)*.

Work done = shaded area = $\frac{1}{2}$ × base × height

$$= \frac{1}{2} \times (20 \times 10^{-3}) \text{ m} \times 500 \text{ N} = \textbf{5 J}$$

4   **Energy** is the capacity, or ability, to do work. The unit of energy is the joule, the same as for work. Energy is expended when work is done.

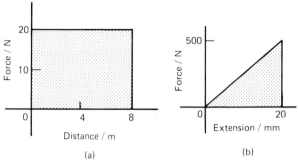

Figure 37.1

5   There are several **forms of energy** and these include:
(i)   Mechanical energy;   (iv)   Chemical energy;
(ii)   Heat or thermal energy;   (v)   Nuclear energy;
(iii) Electrical energy;   (vi)   Light energy;
                                      (vii) Sound energy.

6   Energy may be converted from one form to another. **The principle of conservation of energy** states that the total amount of energy remains the same in such conversions, i.e. energy cannot be created or destroyed. Some examples of energy conversions include:

(i)   Mechanical energy is converted to electrical energy by a generator.
(ii)   Electrical energy is converted to mechanical energy by a motor.
(iii) Heat energy is converted to mechanical energy by a steam engine.
(iv)   Mechanical energy is converted to heat energy by friction.
(v)   Heat energy is converted to electrical energy by a solar cell.
(vi)   Electrical energy is converted to heat energy by an electric fire.
(vii) Heat energy is converted to chemical energy by living plants.
(viii) Chemical energy is converted to heat energy by burning fuels.
(ix)   Heat energy is converted to electrical energy by a thermocouple.
(x)   Chemical energy is converted to electrical energy by batteries.
(xi)   Electrical energy is converted to light energy by a light bulb.

(xii) Sound energy is converted to electrical energy by a microphone.

(xiii) Electrical energy is converted to chemical energy by electrolysis.

7 **Efficiency** is defined as the ratio of the useful output energy to the input energy. The symbol for efficiency is η (Greek letter eta).

$$\text{Hence, } \textbf{efficiency, } \eta = \frac{\textbf{useful output energy}}{\textbf{input energy}}$$

Efficiency has no units and is often stated as a percentage. A perfect machine would have an efficiency of 100%. However, all machines have an efficiency lower than this due to friction and other losses. Thus, if the input energy to a motor is 1000 J and the output energy is 800 J then the efficiency is

$$\frac{800}{1000} \times 100\%, \text{ i.e. } 80\%.$$

8 **Power** is a measure of the rate at which work is done or at which energy is converted from one form to another.

$$\textbf{Power } P = \frac{\textbf{energy used}}{\textbf{time taken}} \left(\text{or } P = \frac{\text{work done}}{\text{time taken}} \right)$$

The unit of power is the **watt**, **W**, where 1 watt is equal to 1 joule per second. The watt is a small unit for many purposes and a larger unit called the kilowatt, kW, is used, where 1 kW = 1000 W. The power output of a motor which does 120 kJ of work is 30 s is thus given by

$$P = \frac{120 \text{ kJ}}{30 \text{ s}} = 4 \text{ kW}.$$

(For electrical power, see page 29.)

9 Since, work done = force × distance

$$\text{Then, power} = \frac{\text{work done}}{\text{time taken}} = \frac{\text{force} \times \text{distance}}{\text{time taken}}$$

$$= \text{force} \times \frac{\text{distance}}{\text{time taken}}$$

$$\text{However, } \frac{\text{distance}}{\text{time taken}} = \text{velocity}.$$

Hence, **power = force × velocity**.

266

Thus, for example, if a lorry is travelling at a constant speed of 72 km/h and the force resisting motion is 800 N, then the tractive power necessary to keep the lorry moving at this speed is given by:

$$\text{power} = \text{force} \times \text{velocity} = (800 \text{ N})\left(\frac{72}{3.6} \text{ m/s}\right) = 16\,000 \,\frac{\text{Nm}}{\text{s}}$$

$$= 16\,000 \text{ J/s} = \textbf{16 kW.}$$

10 (i) Mechanical engineering is concerned principally with two kinds of energy, these being potential energy and kinetic energy.
(ii) **Potential energy** is energy due to the position of a body. The force exerted on a mass of m kg is mg N (where $g = 9.81$ N/kg, the earth's gravitational field). When the mass is lifted vertically through a height h m above some datum level, the work done is given by: force × distance $= (mg)(h)$ J. This work done is stored as potential energy in the mass.

Hence **potential energy** $= \textbf{\textit{mg h}}$ **joules** (the potential energy at the datum level being taken as zero).

(iii) **Kinetic energy** is the energy due to the motion of a body. Suppose a resultant force F acts on an object of mass m originally at rest and accelerates it to a velocity v in a distance s.

Work done = force × distance $= Fs = (ma)(s)$,

where a is the acceleration.

However, $v^2 = 2as$, from which $a = \dfrac{v^2}{2s}$

Hence, work done $= m\left(\dfrac{v^2}{2s}\right)s = \dfrac{1}{2}mv^2$

This energy is called the kinetic energy of the mass m,

i.e. **kinetic energy** $= \dfrac{1}{3}$ ***mv*²** **joules**.

For example, at the instant of striking, a hammer of mass 30 kg has a velocity of 15 m/s. The kinetic energy in the hammer is given by:

$$\text{Kinetic energy} = \frac{1}{2}\textbf{\textit{mv}}_2 = \frac{1}{2} \,(30 \text{ kg})(15 \text{ m/s})^2 = \textbf{3375 J.}$$

11 (i) Energy may be converted from one form to another. The principle of conservation of energy states that the total amount of energy remains the same in such conversions, i.e. energy cannot be created or destroyed.

(ii) In mechanics, the potential energy possessed by a body is frequently converted into kinetic energy, and vice versa. When a mass is falling freely, its potential energy decreases as it loses height, and its kinetic energy increases as its velocity increases. Ignoring air frictional losses, at all times:

potential energy + kinetic energy = a constant.

(iii) If friction is present, then work is done overcoming the resistance due to friction and this is dissipated as heat. Then,

potential energy + = final energy
kinetic energy   + work done overcoming
**   frictional resistance**.

(iv) Kinetic energy is not always conserved in collisions. Collisions in which kinetic energy is conserved (i.e. stays the same) are called **elastic collisions**, and those in which it is not conserved are termed **inelastic collisions**.

Kinetic energy of rotation

12   (i) The tangential velocity v of a particle of mass m moving at an angular velocity ω rad/s at a radius r metres (see *Figure 37.2*) is given by $v = \omega r$ **m/s**.
(ii) The kinetic energy of a particle of mass m is given by:

kinetic energy $= \frac{1}{2} mv^2 = \frac{1}{2} m(\omega r)^2 = \frac{1}{2} m\omega^2 r^2$ **joules**.

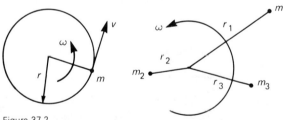

Figure 37.2

Figure 37.3

(iii) The total kinetic energy of a system of masses rotating at different radii about a fixed axis but with the same angular velocity ω, as shown in *Figure 37.3*, is given by:

$$\text{total kinetic energy} = \frac{1}{2} m_1\omega^2 r_1^2 + \frac{1}{2} m_2\omega^2 r_2^2 + \frac{1}{2} m_3\omega^2 r_3^2$$

$$= (m_1 r_1^2 + m_2 r_2^2 + m_3 r_3^2)\, \frac{\omega^2}{2}$$

In general, this may be written as:

total kinetic energy $= (\sum mr^2)\, \dfrac{\omega^2}{2} = I\, \dfrac{\omega^2}{2}$,

where $I\ (= \sum mr^2)$ is called the **moment of inertia** of the system about the axis of rotation.

The moment of inertia of a system is a measure of the amount of work done to give the system an angular velocity of ω rad/s, or the amount of work which can be done by a system turning at ω rad/s.

In general, total kinetic energy $= I\, \dfrac{\omega^2}{2} = Mk^2\, \dfrac{\omega^2}{2}$,

where $M\ (= \sum m)$ is the total mass and k is called the **radius of gyration** of the system for the given axis.

If all of the mass were concentrated at the **radius of gyration** it would give the same moment of inertia as the actual system.

Flywheels

13   The function of a flywheel is to restrict fluctuations of speed by absorbing and releasing large quantities of kinetic energy for small speed variations.

To do this they require large moments of inertia and to avoid excessive mass they need to have radii of gyration as large as possible. Most of the mass of a flywheel is usually in its rim.

38 Torque

1   When two equal forces act on a body as shown in *Figure 38.1*, they cause the body to rotate, and the system of forces is called a **couple**.

2   The turning moment of a couple is called a **torque**, T. In *Figure 38.1*, torque = magnitude of either force × perpendicular distance between the forces, i.e.

$$T = Fd.$$

The unit of torque is the **newton metre**, **Nm**.

3   When a force F newtons is applied at a radius r metres from the axis of, say, a nut to be turned by a spanner, the torque T applied to the nut is given by:

$$T = Fr \textbf{ Nm}$$

4   *Figure 38.2(a)* shows a pulley wheel of radius r metres attached to a shaft and a force F newtons applied to the rim at point P. *Figure 38.2(b)* shows the pulley wheel having turned through an angle θ radians as a result of the force F being applied.

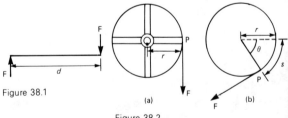

Figure 38.1

(a)

(b)

Figure 38.2

The force moves through a distance s, where arc length $s = r\theta$.

Work done = force × distance moved by force
$$= F \times r\theta = Fr\theta \text{ Nm} = Fr\theta \text{ J}.$$

However, Fr is the torque T,

Hence, **work done** $= T\theta$ **joules**.

5   Power $= \dfrac{\text{work done}}{\text{time taken}}$

$$= \dfrac{T\theta}{\text{time taken}}, \text{ for a constant torque, } T.$$

However, $\dfrac{\text{angle } \theta}{\text{time taken}} = \text{angular velocity, } \omega \text{ rad/s.}$

Hence, **power, $P = T\omega$ watts.**

Angular velocity, $\omega = 2\pi n$ rad/s, where n is the speed in rev/s.

Hence, **power, $P = 2\pi n T$ watts.**

Thus the torque developed by a motor whose spindle is rotating at 1000 rev/min and developing a power of 2.50 kW is given by:

$P = 2\pi n T$, from which,

$$\text{torque } T = \dfrac{P}{2\pi n} = \dfrac{2500}{2\pi\left(\dfrac{1000}{60}\right)}$$

$$= \textbf{23.87 Nm.}$$

6   From para. 4, work done $= T\theta$, and if this work is available to increase the kinetic energy of a rotating body of moment of inertia I, then

$$T\theta = I\left(\dfrac{\omega_2^2 - \omega_1^2}{2}\right)$$

where ω_1 and ω_2 are the initial and final angular velocities,

i.e. $T\theta = I\left(\dfrac{\omega_2 + \omega_1}{2}\right)(\omega_2 - \omega_1)$

However, $\left(\dfrac{\omega_2 + \omega_1}{2}\right)$ is the mean angular velocity, i.e. $\dfrac{\theta}{t}$,

where t is the time and $(\omega_2 - \omega_1)$ is the change in angular velocity, i.e. αt, where α is the angular acceleration.

Hence, $T\theta = I\left(\dfrac{\theta}{t}\right)(\alpha t)$

from which,

$$\boxed{\textbf{torque } T = I\alpha \text{ ,}}$$

where

> I is the moment of inertia in kg m^2,
> α is the angular acceleration in rad/s^2 and
> T is the torque in Nm.

Thus if a shaft system has a moment of inertia of 37.5 kg m^2, the torque required to give it an angular acceleration of 5.0 rad/s^2 is given in:

> torque $T = I\alpha = (37.5)(5.0) = \textbf{187.5 Nm}$.

Power transmission by belt drives

7 (i) A common, and simple method of **transmitting power** from one shaft to another is by means of a **belt** passing over pulley wheels which are keyed to the shafts, as shown in *Figure 38.3*. Typical applications include an electric motor driving a line of shafting and an engine driving a rotating saw.

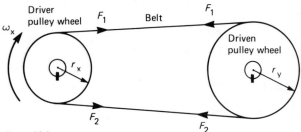

Figure 38.3

(ii) For a belt to transmit power between two pulleys there must be a difference in tensions in the belt on either side of the driving and driven pulleys. For the direction of rotation shown in *Figure 38.3*, $F_2 > F_1$.

The torque T available at the driving wheel to do work is given by:

$$T = (F_2 - F_1)r_x \textbf{ Nm},$$

and the available power P is given by:

$$P = T\omega = (F_2 - F_1)r_x\omega_x \textbf{ watts}.$$

(iii) From para. 12(i), page 268, the linear velocity of a point on the driver wheel, $v_x = r_x \omega_x$. Similarly, the linear velocity of a point on the driven wheel, $v_y = r_y \omega_y$.

Assuming no slipping, $v_x = v_y$

i.e. $r_x \omega_x = r_y \omega_y$

Hence, $r_x(2\pi n_x) = r_y(2\pi n_y)$

from which, $\dfrac{r_x}{r_y} = \dfrac{n_y}{n_x}$

(iv) Efficiency $= \dfrac{\text{useful work output}}{\text{energy input}} \times 100\%$ or

efficiency $= \dfrac{\text{power output}}{\text{power input}} \times 100\%$.

For example, a 15 kW motor is driving a shaft at 1150 rev/min by means of pulley wheels and a belt. The tensions in the belt on each side of the driver pulley are 400 N and 50 N and the diameters of the driver and driven pulley wheel are 500 mm and 750 mm respectively. The power output from the motor is given by:

power output $= (F_2 - F_1) r_x \omega_x$

$$= (400 - 50)\left(\frac{500}{2} \times 10^{-3}\right)\left(\frac{1150 \times 2\pi}{60}\right)$$

$$= 10.54 \text{ kW}.$$

Hence, the efficiency of the motor $= \dfrac{\text{power output}}{\text{power input}}$

$$= \frac{10.54}{15} \times 100\%$$

$$= \mathbf{70.27\%}.$$

The speed of the driven pulley is obtained from

$$\frac{r_x}{r_y} = \frac{n_y}{n_x}$$

i.e. speed of driven pulley wheel,

$$n_y = \frac{n_x r_x}{r_y} = \frac{(1150)(0.25)}{\left(\dfrac{0.75}{2}\right)} = \mathbf{767 \ rev/min}.$$

8   (i) The ratio of the tensions for a flat belt **when the belt is on the point of slipping** is

$$\frac{T_1}{T_2} = e^{\mu\theta}$$

where

μ is the coefficient of friction between belt and pulley,
θ is the angle of lap, in radians (see *Figure 38.4*), and
e is the exponent 2.718.

(ii) For a vee belt as in *Figure 38.5*, the ratio is

$$\frac{T_1}{T_2} = e^{\mu\theta/\sin \alpha}$$

where α is the half angle of the groove and of the belt.

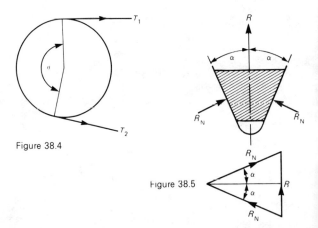

Figure 38.4

Figure 38.5

This gives a much larger ratio than for the flat belt. The V-belt is jammed into its groove and is less likely to slip. Referring to *Figure 38.5*, the force of fraction on each side is μR_N where R_N is the normal (perpendicular) reaction on each side. The triangle of forces shows that $R_N = \dfrac{R/2}{\sin \alpha}$ where R is the resultant reaction.

The force of friction giving rise to the difference between T_1 and T_2 is therefore

$$\mu R_N \times 2 = \left(\frac{\mu}{\sin \alpha}\right) R$$

The corresponding force of friction for a flat belt is μR. Comparing the forces of friction for flat and V-belts it can be said that the V-belt is equivalent to a flat belt with a coefficient of friction given by $\mu/(\sin \alpha)$.

39 Simple machines

1 A machine is a device which can change the magnitude or
line of action, or both magnitude and line of action of a force. A
simple machine usually amplifies an input force, called the **effort**,
to give a larger output force, called the **load**. Some typical
examples of simple machines include pulley systems, screw-jacks,
gear systems and lever systems.

2 The **force ratio** or **mechanical advantage** is defined as the
ratio of load to effort, i.e.

$$\text{force ratio} = \frac{\text{load}}{\text{effort}} \tag{1}$$

Since both load and effort are measured in newtons, force ratio is a
ratio of the same units and thus is a dimensionless quantity.

3 The **movement ratio** or **velocity ratio** is defined as the
ratio of the distance moved by the effort to the distance moved by
the load, i.e.

$$\text{movement ratio} = \frac{\text{distance moved by the effort}}{\text{distance moved by the load}} \tag{2}$$

Since the numerator and denominator are both measured in
metres, movement ratio is a ratio of the same units and thus is a
dimensionless quantity.

4 (i) The **efficiency of a simple machine** is defined as the
ratio of the force ratio to the movement ratio, i.e.

$$\text{efficiency} = \frac{\text{force ratio}}{\text{movement ratio}}$$

Since the numerator and denominator are both dimensionless
quantities, efficiency is a dimensionless quantity. It is usually
expressed as a percentage, thus:

$$\text{efficiency} = \frac{\text{force ratio}}{\text{movement ratio}} \times 100\% \tag{3}$$

(ii) Due to the effects of friction and inertia associated with the
movement of any object, some of the input energy to a machine is

converted into heat and losses occur. Since losses occur, the energy output of a machine is less than the energy input, thus the mechanical effieiency of any machine cannot reach 100%. For example, a simple machine raises a load of 160 kg through a distance of 1.6 m. The effort applied to the machine is 200 N and moves through a distance of 16 m.

Thus, the force ratio$= \dfrac{\text{load}}{\text{effort}} = \dfrac{160 \text{ kg}}{200 \text{ N}} = \dfrac{160 \times 9.81 \text{ N}}{200 \text{ N}} = \textbf{7.85}$,

the movement ratio$= \dfrac{\text{distance moved by effort}}{\text{distance moved by load}}$

$$= \dfrac{16 \text{ m}}{1.6 \text{ m}} = \textbf{10}$$

and, the effieiency$= \dfrac{\text{force ratio}}{\text{movement ratio}} \times 100\%$

$$\dfrac{7.85}{10} \times 100\% = \textbf{78.5\%}.$$

(iii) For simple machines, the relationship between effort and load is of the form: $F_e = aF_1 + b$, where F_e is the effort, F_1 is the load and a and b are constants. From equation (1)

force ratio$= \dfrac{\text{load}}{\text{effort}} = \dfrac{F_1}{F_e} = \dfrac{F_e}{aF_1 + b}$

Dividing both numerator and denominator by F_1 gives:

$$\dfrac{F_1}{aF_1 + b} = \dfrac{1}{a + \dfrac{b}{F_1}}$$

When the load is large, F_1 is large and b/F_1 is small compared with a. The force ratio then becomes approximately equal to $1/a$ and is called the **limiting force ratio**.

The **limiting efficiency** of a simple machine is defined as the ratio of the limiting force ratio to the movement ratio, i.e.

limiting efficiency$= \dfrac{1}{a} \text{ (movement ratio)}$

$$= \dfrac{1}{a \times \text{movement ratio}} \times 100\%,$$

where a is the constant for the law of the machine: $F_e = aF_1 + b$.

Due to friction and inertia, the limiting efficiency of simple machines is usually well below 100%. For example, in a test on a simple machine, the effort-load graph was a straight line of the form $F_e = aF_1 + b$. Two values lying on the graph were at $F_e = 10$ N, $F_1 = 30$ N and at $F_e = 74$ N, $F_1 = 350$ N. The movement ratio of the machine was 17.

The equation $F_e = aF_1 + b$ is of the form $y = mx + c$, where m is the gradient of the graph. The slope of the line passing through points (x_1, y_1) and (x_2, y_2) of the graph $y = mx + c$ is given by:

$$m = \frac{y_2 - y_1}{x_2 - x_1}$$

Thus for $F_e = aF_1 + b$, the slope a is given by:

$$a = \frac{74 - 10}{350 - 30} = \frac{64}{320} = 0.2$$

Hence the limiting force ratio $= \frac{1}{a} = \frac{1}{0.2} = \mathbf{5}$

The limiting efficiency $= \dfrac{1}{a \times \text{movement ratio}} \times 100\%$

$$= \frac{1}{0.2 \times 17} \times 100\% = \mathbf{29.4\%}$$

5 A pulley system is a simple machine. A single-pulley system, shown in *Figure 39.1(a)*, changes the line of action of the effort, but does not change the magnitude of the force.

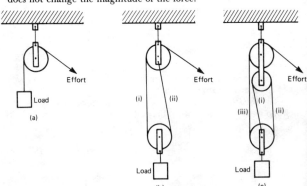

Figure 39.1

278

A two-pulley system, shown in *Figure 39.1(b)*, changes both the line of action and the magnitude of the force. Theoretically, each of the ropes marked (i) and (ii) share the load equally, thus the theoretical effort is only half of the load, i.e. the theoretical force ratio is 2. In practice the actual force ratio is less than 2 due to losses.

A three-pulley system is shown in *Figure 39.1(c)*. Each of the ropes marked (i), (ii) and (iii) carry one-third of the load, thus the theoretical force ratio is 3. In general, for a multiple pulley system having a total of n pulleys, the theoretical force ratio is n. Since the theoretical efficiency of a pulley system (neglecting losses) is 100% and since from equation (3):

$$\text{efficiency} = \frac{\text{force ratio}}{\text{movement ratio}} \times 100\%,$$

it follows that when the force ratio is n,

$$100 = \frac{n}{\text{movement ratio}} \times 100,$$

that is the movement ratio is also n.

For example, a load of 80 kg is lifted by a three-pulley system and the applied effort is 392 N.

$$\text{The force ratio} = \frac{\text{load}}{\text{effort}} = \frac{80 \times 9.81}{392} = \mathbf{2}$$

The movement ratio = 3 (since it is a three-pulley system)

$$\text{Thus the efficiency} = \frac{\text{force ratio}}{\text{movement ratio}} \times 100\%$$

$$= \frac{2}{3} \times 100\% = \mathbf{66.67\%}$$

6   A **simple screw-jack** is shown in *Figure 39.2* and is a simple machine since it changes both the magnitude and the line of action of a force.

The screw of the table of the jack is located in a fixed nut in the body of the jack. As the table is rotated by means of a bar, it raises or lowers a load placed on the table. For a single-start thread, as shown, for one complete revolution of the table, the effort moves through a distance $2\pi r$ and the load moves through a distance equal to the lead of the screw, say l. Thus:

$$\text{the movement ratio} = \frac{2\pi r}{l} \tag{4}$$

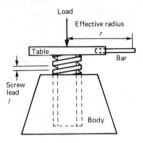

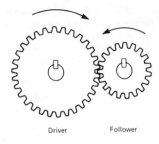

Figure 39.2   Figure 39.3

For example, a screw jack is used to support the axle of a car, the load on it being 2.4 kN. The screw jack has an effort arm of effective radius 200 mm and a single-start square thread having a lead of 5 mm. If an effort of 60 N is required to riase the car axle:

$$\text{Force ratio} = \frac{\text{load}}{\text{effort}} = \frac{2400 \text{ N}}{60 \text{ N}} = \mathbf{40}$$

$$\text{Movement ratio} = \frac{2\pi r}{l} = \frac{2\pi \; 200 \text{ mm}}{5 \text{ mm}} = \mathbf{251.3}$$

$$\text{Hence, efficiency} = \frac{40}{251.3} \times 100\% = \mathbf{15.9\%}.$$

7   (i) A **simple gear train** is used to transmit rotary motion and can change both the magnitude and the line of action of a force, hence is a simple machine. The gear train shown in *Figure 39.3* consists of **spur gears** and has an effort applied to one gear, called the **driver** and a load applied to the other gear, called the **follower**.

(ii) In such a system, the teeth on the wheels are so spaced that they exactly fill the circumference with a whole number of identical teeth, and the teeth on the driver and follower mesh without interference. Under these conditions, the number of teeth on the driver and follower are in direct proportion to the circumference of these wheels, i.e.

$$\frac{\textbf{number of teeth on driver}}{\textbf{number of teeth on follower}}$$

$$= \frac{\textbf{circumference of driver}}{\textbf{circumference of follower}} \quad (5)$$

(iii) If there are, say, 40 teeth on the driver and 20 teeth on the follower then the follower makes two revolutions for each revolution of the driver. In general

$$\frac{\text{the number of revolutions made by the driver}}{\text{the number of revolutions made by the follower}}$$

$$=\frac{\text{the number of teeth on the follower}}{\text{the number of teeth on the driver}} \quad (6)$$

It follows from equation (6) that the speeds of the wheels in a gear train are inversely proportional to the number of teeth.

(iv) The ratio of the speed of the driver wheel to that of the follower is the movement ratio, i.e.

$$\textbf{movement ratio} = \frac{\textbf{speed of driver}}{\textbf{speed of follower}} = \frac{\textbf{teeth on follower}}{\textbf{teeth on driver}} \quad (7)$$

(v) When the same direction of rotation is required on both the driver and the follower an **idler wheel** is used as shown in *Figure 39.4*. Let the driver, idler and follower be A, B and C respectively,

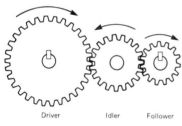

Driver   Idler   Follower

Figure 39.4

and let N be the speed of rotation and T be the number of teeth. Then from equation (7),

$$\frac{N_B}{N_A}=\frac{T_A}{T_B} \quad \text{and} \quad \frac{N_C}{N_B}=\frac{T_B}{T_C}$$

Thus $\dfrac{\text{speed of A}}{\text{speed of C}}=\dfrac{N_A}{N_C}=\dfrac{N_B\dfrac{T_B}{T_A}}{N_B\dfrac{T_B}{T_C}}=\dfrac{T_B}{T_A}\times\dfrac{T_C}{T_B}=\dfrac{T_C}{T_A}$

This shows that the movement ratio is independent of the idler, only the direction of the follower being altered.

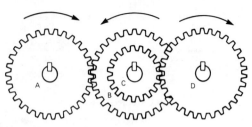

Figure 39.5

(vi) A compound gear train is shown in *Figure 39.5*, in which gear wheels B and C are fixed to the same shaft and hence $N_B = N_C$. From equation (7),

$$\frac{N_A}{N_B} = \frac{T_B}{T_A} \text{ i.e. } N_B = N_A \times \frac{T_A}{T_B}$$

Also, $\frac{N_D}{N_C} = \frac{T_C}{T_D}$, i.e. $N_D = N_C \times \frac{T_C}{T_D}$. But $N_B = N_C$,

hence, $N_D = N_A \times \dfrac{T_A}{T_B} \times \dfrac{T_C}{T_D}$ \hfill (8)

For compound gear trains having say P gear wheels,

$$N_P = N_A \times \frac{T_A}{T_B} \times \frac{T_C}{T_D} \times \frac{T_E}{T_F} \ldots \times \frac{T_O}{T_P} \tag{9}$$

from which

movement ratio $= \dfrac{N_A}{N_P} = \dfrac{T_B}{T_A} \times \dfrac{T_D}{T_C} \times \ldots \times \dfrac{T_P}{T_O}$

For example, a compound gear train consists of a driver gear A having 40 teeth, engaging with gear B, having 160 teeth. Attached to the same shaft as B, gear C has 48 teeth and meshes with gear D on the output shaft having 96 teeth.
Thus from equation (8),

movement ratio $= \dfrac{\text{speed of A}}{\text{speed of D}} = \dfrac{T_B}{T_A} \times \dfrac{T_D}{T_C}$

$$= \frac{160}{40} \times \frac{96}{48} = \mathbf{8}$$

If the force ratio is, say, 6, then the efficiency is $\frac{6}{8} \times 100\% = \mathbf{75\%}$.

8   A **lever** can alter both the magnitude and the line of action

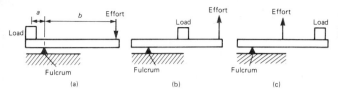

Figure 39.6

of a force and is thus classed as a simple machine. There are three types or orders of levers, as shown in *Figure 39.6*.

(i) A lever of the first order has the fulcrum placed between the effort and the load, as shown in *Figure 39.6(a)*.

(ii) A lever of the second order has the load placed between the effort and fulcrum, as shown in *Figure 39.6(b)*.

(iii) A lever of the third order has the effort applied between the load and the fulcrum, as shown in *Figure 39.6(c)*.

Problems on levers can largely be solved by applying the principle of moments (see page 256). Thus for the lever shown in *Figure 39.6(a)*, when the lever is in equilibrium

anticlockwise moment = clockwise moment

i.e. $a \times F_1 = b \times F_e$

Thus **force ratio** $= \dfrac{F_1}{F_e} = \dfrac{b}{a}$

$$= \frac{\textbf{distance of effort from fulcrum}}{\textbf{distance of load from fulcrum}} \qquad (10)$$

For example, the load on a first-order lever is 1.2 kN, the distance between the fulcrum and load is 0.5 m and the distance between the fulcrum and effort is 1.5 m. Thus, at equilibrium,

anticlockwise moment = clockwise moment

i.e. $1200 \times 0.5 = \text{effort} \times 1.5$

from which, $\quad \text{effort} = \dfrac{1200 \times 0.5}{1.5} = \textbf{400 N}$

Force ratio $= \dfrac{F_1}{F_e} = \dfrac{1200}{400} = \textbf{3}$ $\left(\text{or force ratio} = \dfrac{b}{a} = \dfrac{1.5}{0.5} = 3\right)$

This result shows that to lift a load of say 300 N, an effort of 100 N is required.

40 Heat energy

1   (i) **Heat** is a form of energy and is measured in joules.
(ii) **Temperature** is the degree of hotness or coldness of a
substance.
Heat and temperature are thus **not** the same thing. For example,
twice the heat energy is needed to boil a full container of water
than half a container — that is, different amounts of heat energy
are needed to cause an equal rise in the temperature of different
amounts of the same substance.

2   Temperature is measured either (i) on the **Celsius** (°C)
scale (formerly Centigrade), where the temperature at which ice
melts, i.e. the freezing point of water, is taken as 0°C and the point
at which water boils under normal atmospheric pressure is taken as
100°C, or (ii) on the **thermodynamic scale**, in which the unit of
temperature is the kelvin (K). The kelvin scale uses the same
temperature interval as the Celsius scale but as its zero takes the
'absolute zero of temperature' which is at about 273°C.

Hence kelvin temperature = degree Celsius + 273

i.e. $\mathbf{K} = (\mathbf{°C}) + \mathbf{273}$

Thus, for example,

    0°C = 273 K, 25°C = 298 K and 100°C = 373 K.

3   A **thermometer** is an instrument which measures
temperature. Any substance which possesses one or more properties
which vary with temperature can be used to measure temperature.
These properties include changes in length, area or volume,
electrical resistance or in colour. Examples of temperature
measuring devices include:
(i) **liquid-in-glass thermometer**, which uses the expansion of a
liquid with increase in temperature as its principle of operation,
(ii) **thermocouples**, which use the e.m.f. set up when the
junction of two dissimilar metals is heated,
(iii) **resistance thermometer**, which uses the change in
electrical resistance caused by temperature change, and
(iv) **pyrometers**, which are devices for measuring very high

temperatures, using the principle that all substances emit radiant energy when hot, the rate of emission depending on their temperature.

(See Chapter 42, page 295).

4   (i) The **specific heat capacity** of a substance is the quantity of heat energy required to raise the temperature of 1 kg of the substance by 1°C.

(ii) The symbol used for specific heat capacity is c and the units are J/(kg °C) or J/(kg K). (Note that these units may also be written as J kg^{-1} °C^{-1} or J kg^{-1} K^{-1}.)

(iii) Some typical values of specific heat capacity for the range of temperature 0°C to 100°C include:

water	4190 J/(kg °C),	ice	2100 J/(kg °C)
aluminium	950 J/(kg °C),	copper	390 J/(kg °C)
iron	500 J/(kg °C)	lead	130 J/(kg °C).

Hence

to raise the temperature of 1 kg of iron by 1°C requires 500 J of energy,

to raise the temperature of 5 kg of iron by 1°C requires (500×5) J of energy, and

to raise the temperature of 5 kg of iron by 40°C requires $(500 \times 5 \times 40)$ J of energy, i.e. 100 kJ.

In general, the quantity of heat energy, Q, required to raise a mass m kg of a substance with a specific heat capacity c J/(kg °C) from temperature t_1 °C to t_2 °C is given by:

$$Q = mc(t_2 - t_1) \quad \textbf{joules.}$$

5   A material may exist in any one of three states — solid, liquid or gas. If heat is supplied at a constant rate to some ice initially at, say, -30°C, its temperature rises as shown in *Figure 40.1*. Initially the temperature increases from -30°C to 0°C as shown by the line AB. It then remains constant at 0°C for the time BC required for the ice to melt into water. When melting commences the energy gained by continual heating is offset by the energy required for the change of state and the temperature remains constant even though heating is continued. When the ice is completely melted to water, continual heating raises the temperature to 100°C, as shown by CD in *Figure 40.1*. The water then begins to boil and the temperature again remains constant at 100°C, shown as DE, until all the water has vaporised. Continual heating raises the temperature of the steam as shown by EF in the region where the steam is termed superheated. Changes of state from solid to liquid or liquid to gas occur without change of

temperature and such changes are reversible processes. When heat energy flows to or from a substance and causes a change of temperature, such as between A and B, between C and D and between E and F in *Figure 40.1*, it is called **sensible heat** (since it can be 'sensed' by a thermometer).

Heat energy which flows to or from a substance while the temperature remains constant, such as between B and C and between D and E in *Figure 40.1*, is called **latent heat** (latent means concealed or hidden).

6   (i) The **specific latent heat of fusion** is the heat required to change 1 kg of a substance from the solid state to the liquid state (or vice versa) at constant temperature.

(ii) The **specific latent heat of vaporisation** is the heat required to change 1 kg of a substance from a liquid to a gaseous state (or vice versa) at constant temperature.

(iii) The units of the specific latent heats of fusion and vaporisation are J/kg, or more often, kJ/kg, and some typical values are shown below.

	Latent heat of fusion (kJ/kg)	*Melting point (°C)*
Mercury	11.8	− 39
Lead	22	327
Silver	100	957
Ice	335	0
Aluminium	387	660

	Latent heat of vaporisation (kJ/kg)	*Boiling point (°C)*
Oxygen	214	− 183
Mercury	286	357
Ethyl alcohol	857	79
Water	2257	100

(iv) The quantity of heat Q supplied or given out during a change of state is given by:

$$Q = mL$$

where m is the mass in kilograms and L is the specific latent heat.

Thus, for example, the heat required to convert 10 kg of ice at 0°C

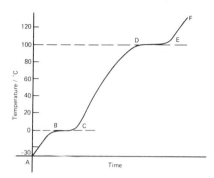

Figure 40.1

to water at 0°C is given by 10 kg × 335 kJ/kg, i.e. 3350 kJ or
3.35 MJ.

Principle of operation of a refrigerator

7   The boiling point of most liquids may be lowered if the
pressure is lowered. In a simple refrigerator a working fluid, such as
ammonia or freon, has the pressure acting on it reduced. The
resulting lowering of the boiling point causes the liquid to vaporise.

In vaporising, the liquid takes in the necessary latent heat
from its surroundings, i.e. the freezer, which thus becomes cooled.
The vapour is immediately removed by a pump to a condenser
which is outside of the cabinet, where it is compressed and
changed back into a liquid, giving out latent heat. The cycle is
repeated when the liquid is pumped back to the freezer to be
vaporised.

Conduction, convection and radiation

8   Heat may be **transferred** from a hot body to a cooler body
by one or more of three methods, these being: (a) by **conduction**,
(b) by **convection**, or (c) by **radiation**.

9   **Conduction** is the transfer of heat energy from one part of a
body to another (or from one body to another) without the
particles of the body moving.

Conduction is associated with solids. For example, if one end
of a metal bar is heated, the other end will become hot by
conduction. Metals and metallic alloys are good conductors of heat
whereas air, wood, plastic, cork, glass and gases are examples of
poor conductors (i.e. heat insulators).

Practical applications of conduction:

(i) A domestic saucepan or dish conducts heat from the source to the contents. Also, since wood and plastic are poor conductors of heat they are used for saucepan handles.

(ii) The metal of a radiator of a central heating system conducts heat from the hot water inside to the air outside.

10 **Convection** is the transfer of heat energy through a substance by the actual movement of the substance itself. Convection occurs in liquids and gases, but not in solids. When heated, a liquid or gas becomes less dense. It then rises and is replaced by a colder liquid or gas and the process repeats. For example, electric kettles and central heating radiators always heat up at the top first.

Examples of convection are:

(i) Natural circulation hot water heating systems depend on the hot water rising by convection to the top of a house and then falling back to the bottom of the house as it cools, releasing the heat energy to warm the house as does so.

(ii) Convection currents cause air to move and therefore affect climate.

(iii) When a radiator heats the air around it, the hot air rises by convection and cold air moves in to take its place.

(iv) A cooling system in a car radiator relies on convection.

(v) Large electrical transformers dissipate waste heat to an oil tank. The heated oil rises by convection to the top, then sinks through cooling fins, losing heat as it does so.

(vi) In a refrigerator, the cooling unit is situated near the top. The air surrounding the cold pipes becomes heavier as it contracts and sinks towards the bottom. Warmer, less dense air is pushed upwards and in turn is cooled. A cold convection current is thus created.

11 **Radiation** is the transfer of heat energy from a hot body to a cooler one by electromagnetic waves. Heat radiation is similar in character to light waves (see Chapter 31) — it travels at the same speed and can pass through a vacuum — except that the frequency of the waves are different. Waves are emitted by a hot body, are transmitted through space (even a vacuum), and are not detected until they fall on to another body. Radiation is reflected from shining, polished surfaces but absorbed by dull, black surfaces.

Practical applications of radiation include:

(i) heat from the sun reaching earth;
(ii) heat felt by a flame;
(iii) cooker grills;

(iv) industrial furnaces;

(v) infra-red space heaters.

Vacuum flask

12   A cross-section of a typical vacuum flask is shown in *Figure 40.2* and is seen to be a double-walled bottle with a vacuum space between them, the whole supported in a protective outer case.

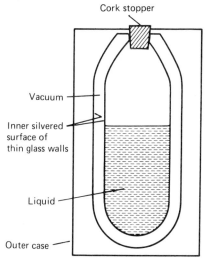

Figure 40.2

Very little heat can be transferred by conduction because of the vacuum space and the cork stopper (cork is a bad conductor of heat). Also, because of the vacuum space, no convection is possible. Radiation is minimised by silvering the two glass surfaces (radiation is reflected off of shining surfaces).

Thus a vacuum flask is an example of prevention of all three types of heat transfer and is therefore able to keep hot liquids hot and cold liquids cold.

Use of insulation in conserving fuel

13   Fuel used for heating a building is becoming increasingly expensive. By the careful use of insulation, heat can be retained in

a building for longer periods and the cost of heating thus minimised.

(i) Since convection causes hot air to rise it is important to insulate the roof space, which is probably the greatest source of heat loss in the home. This can be achieved by laying fibre-glass between the wooden joists in the roof space.

(ii) Glass is a poor conductor of heat. However, large losses can occur through thin panes of glass and such losses can be reduced by using double-glazing. Two sheets of glass, separated by air, are used. Air is a good insulator but the air space must not be too large otherwise convection currents can occur which would carry heat across the space.

(iii) Hot water tanks should be lagged to prevent conduction and convection of heat to the surrounding air.

(iv) Brick, concrete, plaster and wood are all poor conductors of heat. A house is made from two walls with an air gap between them. Air is a poor conductor and trapped air minimises losses through the wall. Heat losses through walls can be prevented almost completely by using cavity wall insulation, i.e. plastic-foam.

14　Besides changing temperature, the effects of supplying heat to a material can involve changes in dimensions, as well as in colour, state and electrical resistance.

Most substances expand when heated and contract when cooled, and there are many practical applications and design implications of thermal movement. (See Chapter 41.)

41 Thermal expansion

1   When heat is applied to most materials, **expansion** occurs in all directions. Conversely, if heat energy is removed from a material (i.e. the material is cooled) **contraction** occurs in all directions. The effects of expansion and contraction each depend on the **change of temperature** of the material.

2   Some practical applications where expansion and contraction of solid materials must be allowed for include:

(i) Overhead electrical transmission lines are hung so that they are slack in summer, otherwise their contraction in winter might snap the conductors or bring down pylons.

(ii) Gaps need to be left in lengths of railway lines to prevent buckling in hot weather.

(iii) Ends of large bridges are often supported on rollers to allow them to expand and contract freely.

(iv) Fitting a metal collar to a shaft or a steel tyre to a wheel is often achieved by first heating them so that they expand, fitting them in position, and then cooling them so that the contraction holds them firmly in place. This is known as a 'shrink-fit'. By a similar method hot rivets are used for joining metal sheets.

(v) The amount of expansion varies with different materials. *Figure 41.1(a)* shows a bimetallic strip at room temperature (i.e. two different strips of metal riveted together).

When heated, brass expands more than steel, and since the two metals are riveted together the bimetallic strip is forced into an arc as shown in *Figure 41.1(b)*. Such a movement can be arranged to make or break an electric circuit and bimetallic strips are used, in particular, in thermostats (which are temperature operated switches) used to control central heating systems, cookers, refrigerators, toasters, irons, hot-water and alarm systems.

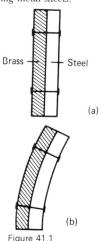

Figure 41.1

(vi) Motor engines use the rapid expansion of heated gases to force a piston to move.

(vii) Designers must predict, and allow for, the expansion of steel pipes in a steam-raising plant so as to avoid damage and consequent danger to health.

3 (i) Water is a liquid which at low temperatures displays an unusual effect. If cooled, contraction occurs until, at about 4°C, the volume is at a minimum. As the temperature is further decreased from 4°C to 0°C expansion occurs, i.e. the volume increases. When ice is formed considerable expansion occurs and it is this expansion which often causes frozen water pipes to burst.

(ii) A practical application of the expansion of a liquid is with thermometers, where the expansion of a liquid, such as mercury or alcohol, is used to measure temperature.

4 (i) The amount by which unit length of a material expands when the temperature is raised one degree is called the **coefficient of linear expansion** of the material and is represented by α (Greek alpha).

(ii) The units of the coefficient of linear expansion are m/(m K), although it is usually quoted as just /K or K^{-1}. For example, copper has a coefficient of linear expansion value of 17×10^{-6} K^{-1}, which means that a 1 m long bar of copper expands by 0.000017 m if its temperature is increased by 1 K (or 1°C). If a 6 m long bar of copper is subjected to a temperature rise of 25 K then the bar will expand by $(6 \times 0.000017 \times 25)$ m, i.e. 0.00255 m or 2.55 mm. (Since the kelvin scale uses the same temperature interval as the Celsius scale, a **change** of temperature of, say, 50°C, is the same as a **change** of temperature of 50 K.)

(iii) If a material, initially of length l_1 and at a temperature t_1 and having a coefficient of linear expansion α, has its temperature increased to t_2, then the new length l_2 of the material is given by:

new length = original length + expansion

i.e. $l_2 = l_1 + l_1 \alpha (t_2 - t_1)$

i.e. $\boxed{l_2 = l_1 [1 + \alpha(t_2 - t_1)]}$ (1)

(iv) Some typical values for the coefficient of linear expansion include:

aluminium	23×10^{-6} K^{-1},	brass	18×10^{-6} K^{-1}
concrete	12×10^{-6} K^{-1},	copper	17×10^{-6} K^{-1}
gold	14×10^{-6} K^{-1},	invar (nickel-steel alloy)	
iron	$11–12 \times 10^{-6}$ K^{-1},		0.9×10^{-6} K^{-1}
steel	$15–16 \times 10^{-6}$ K^{-1},	nylon	100×10^{-6} K^{-1}
zinc	31×10^{-6} K^{-1},	tungsten	4.5×10^{-6} K^{-1}.

For example, the copper tubes in a boiler are 4.20 m long at a temperature of 20°C. Then, when surrounded only by feed water at 10°C, the final length of the tubes l_2 is given by:

$$l_2 = l_1[1 + \alpha(t_2 - t_1)] = 4.20[1 + (17 \times 10^{-6})(10 - 20)]$$
$$= \textbf{4.1993 m},$$

i.e. the tube contracts by 0.7 mm when the temperature decreases from 20°C to 10°C.

When the boiler is operating and the mean temperature of the tubes is 320°C, then the length of the tube l_2 is given by:

$$l_2 = l_1[1 + \alpha(t_2 - t_1)] = 4.20[1 + (17 \times 10^{-6})(320 - 20)]$$
$$= 4.20[1 + 0.0051] = \textbf{4.2214 m},$$

i.e. the tube extends by 21.4 mm when the temperature rises from 20°C to 320°C.

5   (i) The amount by which unit area of a material increases when the temperature is raised by one degree is called the **coefficient of superficial (i.e. area) expansion** and is represented by β (Greek beta).

(ii) If a material having an initial surface area A_1 at temperature t_1 and having a coefficient of superficial expansion β, has its temperature increased to t_2, then the new surface area A_2 of the material is given by:

new surface area = original surface area + increase in area

i.e. $A_2 = A_1 + A_1\beta(t_2 - t_1)$

i.e. $\boxed{A_2 = A_1[1 + \beta(t_2 - t_1)]}$ (2)

(iii) It may be shown that the coefficient of superficial expansion is twice the coefficient of linear expansion, i.e. $\beta = 2\alpha$, to a very close approximation.

6   (i) The amount by which unit volume of a material increases for a one degree rise of temperature is called the **coefficient of cubic (or volumetric) expansion** and is represented by γ (Greek gamma).

(ii) If a material having an initial volume V_1 at temperature t_1 and having a coefficient of cubic expansions γ, has its temperature raised to t_2, then the new volume V_2 of the material is given by:

new volume = initial volume + increase in volume

i.e. $V_2 = V_1 + V_1\gamma(t_2 - t_1)$

i.e. $\boxed{V_2 = V_1[1 + \gamma(t_2 - t_1)]}$ (3)

(iii) It may be shown that the coefficient of cubic expansion is three times the coefficient of linear expansion, i.e. $\gamma = 3\alpha$, to a very close approximation. A liquid has no definite shape and only its cubic or volumetric expansion need be considered. Thus with expansions in liquids, equation (3) is used.

(iv) Some typical values for the coefficient of cubic expansion measured at 20°C (i.e. 293 K) include:

ethyl alcohol	1.1×10^{-3} K^{-1},	mercury	1.82×10^{-4} K^{-1},
paraffin oil	9×10^{-2} K^{-1},	water	2.1×10^{-4} K^{-1}.

The coefficient of cubic expansion γ is only constant over a limited range of temperature.

For example, mercury contained in a thermometer has a volume of 476 mm^3 at 15°C. When the volume is, say, 478 mm^3, then:

$$V_2 = V_1[1 + \gamma(t_2 - t_1)]$$

$$478 = 476[1 + 1.82 \times 10^{-4}(t_2 - 15)]$$

from which, $478 = 476 + (476)(1.82 \times 10^{-4})(t_2 - 15)$

$$\frac{478 - 476}{(476)(1.82 \times 10^{-4})} = t_2 - 15$$

$$23.09 = t_2 - 15$$

Hence when the volume of mercury is 478 mm^3 the temperature is $23.09 + 15 = \textbf{38.09°C}$.

42 Measurement of temperature

1   A change in temperature of a substance can often result in a change in one or more of its physical properties. Thus, although temperature cannot be measured directly, its effects can be measured. Some properties of substances used to determine changes in temperature include changes in dimensions, electrical resistance, state, type and volume of radiation and colour. Temperature measuring devices available are many and varied. The devices described in the following paragraphs are most often used in science and industry.

Liquid-in-glass thermometer

2   **Construction**

   (a) A typical liquid-in-glass thermometer is shown in *Figure 42.1* and consists of a sealed stem of uniform small bore tubing, called a capillary tube, made of glass, with a cylindrical glass bulb formed at one end. The bulb and

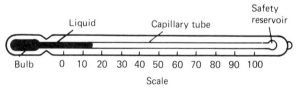

Figure 42.1

part of the stem are filled with a liquid such as mercury or alcohol and the remaining part of the tube is evacuated. A temperature scale is formed by etching graduations on the stem (see (b)). A safety reservoir is usually provided into which the liquid can expand without bursting the glass if the temperature is raised beyond the upper limit of the scale.

Principle of operation

(b) The operation of a liquid-in-glass thermometer depends on the liquid expanding with increase in temperature and contracting with decrease in temperature. The position of the end of the column of liquid in the tube is a measure of the temperature of the liquid in the bulb — shown as 15°C in *Figure 42.1*, which is about room temperature. Two fixed points are needed to calibrate the thermometer, with the interval between these points beting divided into 'degrees'. In the first thermometer, made by Celsius, the fixed points chosen were the temperature of melting ice (0°C), and that of boiling water at standard atmospheric pressure (100°C), in each case the blank stem being marked at the liquid level. The distance between these two points, called the fundamental interval, was divided into 100 equal parts, each equivalent to 1°C, thus forming the scale.

The cylindrical thermometer, with a limited scale around body temperature, the maximum and/or minimum thermometer, recording the maximum day temperature and minimum night temperature, and the Beckman thermometer, which is used only in accurate measurement of temperature change and has no fixed points, are particular types of liquid in glass thermometer which all operate on the same principle.

Advantages

(c) The liquid in glass thermometer is simple in construction, relatively inexpensive, easy to use and portable, and is the most widely used method of temperature measurement having industrial, chemical, clinical and meteorological applications.

Disadvantages

(d) Liquid in glass thermometers tend to be fragile and hence easily broken, can only be used where the liquid column is visible, cannot be used for surface temperature measurements, cannot be read from a distance and are unsuitable for high temperature measurements.

(e) **The use of mercury in a thermometer has many advantages**, for mercury:
  (i) is clearly visible;
  (ii) has a fairly uniform rate of expansion;
  (iii) is readily obtainable in the pure state;

(iv) does not 'wet' the glass; and

(v) is a good conductor of heat.

Mercury has a freezing point of $-39°C$ and cannot be used in a thermometer below this temperature. Its boiling point is $357°C$ but before this temperature is reached some distillation of the mercury occurs if the space above the mercury is a vacuum. To prevent this, and to extend the upper temperature limits to over $500°C$, an inert gas such as nitrogen under pressure is used to fill the remainder of the capillary tube. Alcohol, often dyed red to be seen in the capillary tube, is considerably cheaper than mercury and has a freezing point of $-113°C$, which is considerably lower than for mercury. However it has a low boiling point at about $79°C$.

(f) **Typical errors in liquid in glass thermometers may occur due to:**

(i) the slow cooling rate of glass,

(ii) incorrect positioning of the thermometer,

(iii) a delay in the thermometer becoming steady (i.e. slow response time), and

(iv) non-uniformity of the bore of the capillary tube, which means that equal intervals marked on the stem do not correspond to equal temperature intervals.

Thermocouple

3   **Principle of operation**

(a) At the junction between two different metals, say, copper and constantan, there exists a difference in electrical potential, which varies with the temperature of the junction. This is known as the 'thermoelectric effect'. If the circuit is completed with a second junction at a different temperature, a current will flow round the circuit. This principle is used in the thermocouple. Two different metal conductors having their ends twisted together are shown in *Figure 42.2*. If the two junctions are at different temperatures, a current I flows round the circuit.

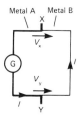

Figure 42.2

The deflection on the galvanometer G depends on the difference in temperature between junction X and Y and is caused by the difference between voltages V_X and V_Y. The higher temperature junction is usually called the 'hot junction' and the lower temperature junction the 'cold junction'. If the cold junction is kept at a constant known temperature, the galvanometer can be calibrated to indicate the temperature of the hot junction directly. The cold junction is then known as the reference junction.

In many instrumentation situations, the measuring instrument needs to be located far from the point at which the measurements are to be made. Extension leads are then used, usually made of the same material as the thermocouple but of smaller gauge. The reference junction is then effectively moved to their ends.

The thermocouple is used by positioning the hot junction where the temperature is required. The meter will indicate the temperature of the hot junction only if the reference junction is at 0°C for:

(temperature of hot junction) = (temperature of the cold junction) + (temperature difference).

In a laboratory the reference junction is often placed in melting ice, but in industry it is often positioned in a thermostatically controlled oven or buried underground where the temperature is constant.

Construction and typical applications

(b) Thermocouple junctions are made by twisting together two wires of dissimilar metals before welding them. The construction of a typical copper-constantan thermocouple for industrial use is shown in *Figure 42.3*.

Apart from the actual junction the two conductors used must be electrically insulated from each other with appropriate insulation and is shown in *Figure 42.3* as twin-holed tubing. The wires and insulation are usually inserted into a sheath for protection from environments in which it might be damaged or corroded. A copper-constantan thermocouple can measure temperature from −250°C up to about 400°C, and is used typically with boiler flue gases, food processing and with sub-zero temperature measurement.

An iron-constantan thermocouple can measure temperature from −200°C to about 850°C, and is used

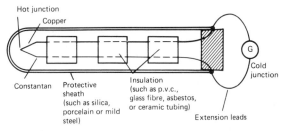

Hot junction
Copper
Constantan
Protective sheath (such as silica, porcelain or mild steel)
Insulation (such as p.v.c., glass fibre, asbestos, or ceramic tubing)
Cold junction
Extension leads
G

Figure 42.3

typically in paper and pulp mills, reheat and annealing furnaces and in chemical reactors. A chromal-alumel thermocouple can measure temperatures from −200°C to about 1100°C and is used typically with blast furnace gases, brick kilns and in glass manufacture.

For the measurement of temperatures above 1100°C radiation pyrometers are usually used. However, thermocouples are available made of platinum-platinum/rhodium capable of measuring temperatures up to 1400°C or tungsten-molybedenum which can measure up to 2600°C.

Advantages of thermocouples over other devices

(c) A thermocouple
    (i) has a very simple relatively inexpensive construction;
    (ii) can be made very small and compact;
    (iii) is robust;
    (iv) is easily replaced if damaged;
    (v) has a small response time;
    (vi) can be used at a distance from the actual measuring instrument and is thus ideal for use with automatic and remote-control systems.

(d) **Sources of error in the thermocouple which are difficult to overcome include:**
    (i) voltage drops in leads and junctions;
    (ii) possible variations in the temperature of the cold junction; and
    (iii) stray thermoelectric effects, which are caused by the addition of further metals into the 'ideal' two metal thermocouple circuit. Additional leads are frequently necessary for extension leads or voltmeter terminal connections.

299

A thermocouple may be used with a battery or mains operated electronic thermometer instead of a millivoltmeter. These devices amplify the small e.m.f.'s from the thermocouple before feeding them to a multi-range voltmeter calibrated directly with temperature scales. They have great accuracy and are almost unaffected by voltage drops in the leads and junctions.

Resistance thermometer

Construction

4 (a) Resistance thermometers are made in a variety of sizes, shapes and forms depending on the application for which they are designed.

A typical resistance thermometer is shown diagrammatically in *Figure 42.4*. The most common metal used for the coil in such thermometers is platinum even

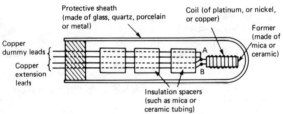

Figure 42.4

though its sensitivity is not as high as other metals such as copper and nickel. However, platinum is a very stable metal and provides reproducible results in a resistance thermometer. A platinum resistance thermometer is often used as a calibrating device. Since platinum is expensive connecting leads of another metal, usually copper, are used with the thermometer to connect it to a measuring circuit. The platinum and the connecting leads are shown joined at A and B in *Figure 42.4*, although sometimes this junction may be made outside of the sheath. However, these leads often come into close contact with the heat source which can introduce errors into the measurements. This may be eliminated by including a pair of identical leads, called dummy leads, which experience the same temperature change as the extension leads.

Principle of operation

(b) With most metals a rise in temperature causes an increase in electrical resistance, and since resistance can be measured accurately this property can be used to measure temperature. If the resistance of a length of wire at 0°C is R_o, and its resistance at θ°C is R_θ,

Then, $R_\theta = R_o(1 + \alpha\theta)$,

where α is the temperature coefficient of resistance of the material.

Rearranging gives: **temperature**, $\theta = \dfrac{R_\theta - R_o}{\alpha R_o}$

Values of R_o and α may be determined experimentally or obtained from existing data. Thus, if R_θ can be measured, temperature θ can be calculated. This is the principle of operation of a resistance thermometer. Although a sensitive ohmmeter can be used to measure R_θ, for more accurate determinations a Wheatstone bridge circuit is used as shown in *Figure 42.5*. This circuit

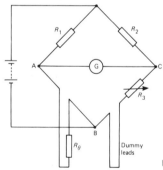

Figure 42.5

compares an unknown resistance R_θ with others of known values, R_1 and R_2 being fixed values and R_3 being variable. Galvanometer G is a sensitive centre-zero microammeter. R_3 is varied until zero deflection is obtained on the galvanometer, i.e. no current flows through G and the bridge is said to be 'balanced'.

At balance: $R_2R_\theta = R_1R_3$, from which

$$R_\theta = \frac{R_1R_3}{R_2},$$

and if R_1 and R_2 are of equal value, then $R_\theta = R_3$.

A resistance thermometer may be connected between points A and B in *Figure 42.5* and its resistance R_θ at any temperature θ accurately measured. Dummy leads are included in arm BC to help eliminate errors caused by the extension leads which are normally necessary in such a thermometer.

(c) Resistance thermometers using a nickel coil are used mainly in the range $-100°C$ to $300°C$, whereas platinum resistance thermometers are capable of measuring with great accuracy temperatures in the range $-200°C$ to about $800°C$. This upper range may be extended up to about $1500°C$ if high melting point materials are used for the sheath and coil construction.

(d) Platinum is commonly used in resistance thermometers since it is chemically inert, i.e. unreactive, resists corrosion and oxidation and has a high melting point of $1769°C$. A disadvantage of platinum is its slow response to temperature variation.

(e) Platinum resistance thermometers may be used as a calibrating device or in such applications as heat treating and annealing processes and it can easily be adopted for use with automatic recording or control systems. Resistance thermometers tend to be fragile and easily damaged especially when subjected to excessive vibration or shock.

Thermistors

5   A thermistor is a semiconducting material — such as mixtures of oxides of copper, mangenese, cobalt, etc. — in the form of a fused bead connected to two leads. As its temperature is increased its resistance rapidly decreases.

Typical resistance/temperature curves for a thermistor and common metals is shown in *Figure 42.6*. The resistance of a typical thermistor can vary from $400\ \Omega$ at $0°C$ to $100\ \Omega$ at $140°C$.

The main advantages of a thermistor are its high sensitivity and small size. It provides an inexpensive method of measuring and detecting small changes in temperature.

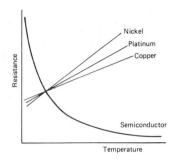

Figure 42.6

Pyrometers

6 (a) A pyrometer is a device for measuring very high
temperatures and uses the principle that all substances
emit radiant energy when hot, the rate of emission
depending on their temperature. The measurement of
thermal radiation is therefore a convenient method of
determining the temperature of hot sources and is
particularly useful in industrial processes. There are two
main types of pyrometers, these being the total radiation
pyrometer and the optical pyrometer.

Pyrometers are very convenient instruments since
they can be used at a safe and comfortable distance from
the hot source. Thus applications of pyrometers are found
in measuring the temperature of molten metals, the
interiors of furnaces or the interiors of volcanos. Total
radiation pyrometers can also be used in conjunction
with devices which record and control temperature
continuously.

Total radiation pyrometer

(b) A typical arrangement of a total radiation pyrometer is
shown in *Figure 42.7*. Radiant energy from a hot source,
such as a furnace, is focussed on to the hot junction of a
thermocouple after reflection from a concave mirror. The
temperature rise recorded by the thermocouple depends
on the amount of radiant energy received, which in turn
depends on the temperature of the hot source. The
galvanometer G shown connected to the thermocouple
records the current which results from the e.m.f.
developed and may be calibrated to give a direct reading

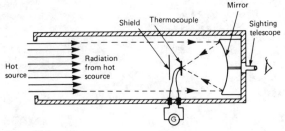

Figure 42.7

of the temperature of the hot source. The thermocouple is protected from direct radiation by a shield as shown and the hot source may be viewed through the sighting telescope. For greater sensitivity, a thermopile may be used, a thermopile being a number of thermocouples connected in series. Total radiation pyrometers are used to measure temperature in the range 700°C to 2000°C.

Optical pyrometer

(c) When the temperature of an object is raised sufficiently two visual effects occur. These are that the object appears brighter and that there is a change in colour of the light emitted. These effects are used in the optical pyrometer where a comparison or matching is made between the brightness of the glowing hot source and the light from a filament of known temperature.

The most frequently used optical pyrometer is the disappearing filament pyrometer and a typical arrangement is shown in *Figure 42.8*. A filament lamp is

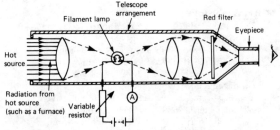

Figure 42.8

built into a telescope arrangement which receives radiation from a hot source, an image of which is seen through an eyepiece.

A red filter is incorporated as a protection to the eye. The current flowing through the lamp is controlled by a variable resistor. As the current is increased the temperature of the filament increases and its colour changes. When viewed through the eyepiece the filament of the lamp appears superimposed on the image of the radiant energy from the hot source. The current is varied until the filament glows as brightly as the background. It will then merge into the background and seem to disappear. The current required to achieve this is a measure of the temperature of the hot source and the ammeter can be calibrated to read the temperature directly. Optical pyrometers may be used to measure temperatures up to, and even in excess of, 3000°C.

Advantages of pyrometers

(d) (i) There is no practical limit to the temperature that a pyrometer can measure.

(ii) A pyrometer need not be brought directly into the hot zone and so is free from the effect of heat and chemical attack that can often cause other measuring devices to deteriorate in use.

(iii) Very fast rates of changes of temperature can be followed by a pyrometer.

(iv) The temperature of moving bodies can be measured.

(v) The lens system makes the pyrometer virtually independent of its distance from the source.

Disadvantages of pyrometers

(e) (i) A pyrometer is often more expensive than other temperature measuring devices.

(ii) A direct view of the heat process is necessary.

(iii) Manual adjustment is necessary.

(iv) A reasonable amount of skill and care is required in calibrating and using a pyrometer. For each new measuring situation the pyrometer must be re-calibrated.

(v) The temperature of the surroundings may affect the reading of the pyrometer and such errors are difficult to eliminate.

7 (a) **Temperature indicating paints** contain coloured substances which change their colour when heated to certain temperatures. This change is usually due to

chemical decomposition, such as loss of water, in which the change in colour of the paint after having reached the particular temperature will be a permanent one. However in some types, the original colour returns after cooling. Temperature indicating paints are used where the temperature of inaccessible parts of apparatus and machines is required. They are particularly useful in heat-treatment processes where the temperature of the component needs to be known before a quenching operation. There are several such paints available and most have only a small temperature range so that different paints have to be used for different temperatures. The usual range of temperatures covered by these paints is from about 30°C to 700°C.

(b) **Temperature sensitive crayons** consist of fusible solids compressed into the form of a stick. The melting point of such crayons is used to determine when a given temperature has been reached. The crayons are simple to use but indicate a single temperature only, i.e. its melting point temperature. There are over a hundred different crayons available, each covering a particular range of temperature. Crayons are available for temperatures within the range of 50°C to 1400°C. Such crayons are used in metallurgical applications such as preheating before welding, hardening, annealing or tempering, or in monitoring the temperature of critical parts of machines or for checking mould temperatures in the rubber and plastics industry.

8   **Bimetallic thermometers** depend on the expansion of metal strips which operate an indicating pointer. Two thin metal strips of differing thermal expansion are welded or riveted together and the curvature of the bimetallic strip changes with temperature change. For greater sensitivity the strips may be coiled into a flat spiral or helix, one end being fixed and the other being made to rotate a pointer over a scale. Bimetallic thermometers are useful for alarm and overtemperature applications where extreme accuracy is not essential. If the whole is placed in a sheath, protection from corrosive environments is achieved but with a reduction in response characteristics. The normal upper limit of temperature measurement by this thermometer is about 200°C, although with special metals the range can be extended to about 400°C.

9   The **mercury in steel thermometer** is an extension of the principle of the mercury in glass thermometer. Mercury in a steel bulb expands via a small bore capillary tube into a pressure indicating device, say, a Bourdon gauge, the position of the pointer

indicating the amount of expansion and thus the temperature. The advantages of this instrument are that it is robust, and, by increasing the length of the capillary tube, the gauge can be placed some distance from the bulb and can thus be used to monitor temperatures in positions which are inaccessible to the liquid in glass thermometer. Such thermometers may be used to measure temperatures up to 600°C.

10   The **gas thermometer** consists of a flexible U-tube of mercury connected by a capillary tube to a vessel containing gas. The change in the volume of a fixed mass of gas at constant pressure, or the change in pressure of a fixed mass of gas at constant volume, may be used to measure temperature. This thermometer is cumbersome and rarely used to measure temperature directly. It is often used as a standard with which to calibrate other types of thermometer. With pure hydrogen the range of the instrument extends from $-240°C$ to $1500°C$ and measurements can be made with extreme accuracy.

43 Pressure in fluids

1   The **pressure** acting on a surface is defined as the
perpendicular force per unit area of surface. The unit of pressure is
the **pascal**, (**Pa**), where 1 pascal is equal to 1 newton per square
metre. Thus

$$\boxed{\text{pressure, } p = \frac{F}{A} \text{ pascals}}$$

where F is the force in newtons acting at right angles to a surface
of area A square metres.

When a force of 20 N acts uniformly over, and perpendicular
to an area of 4 m^2, then the pressure on the area, p, is given by:

$$p = \frac{20 \text{ N}}{4 \text{ m}^2} = 5 \text{ Pa}$$

2   A **fluid** is either a liquid or a gas and there are four basic
factors governing the pressure within fluids.

   (a) The pressure at a given depth in a fluid is equal in all
        directions, see *Figure 43.1(a)*.
   (b) The pressure at a given depth in a fluid is independent of
        the shape of the container in which the fluid is held. In
        Figure 43.1(b), the pressure at X is the same as the
        pressure at Y.
   (c) Pressure acts at right angles to the surface containing the
        fluid. In *Figure 43.1(c)*, the pressure at points A to F all
        act at right angles to the container.
   (d) When a pressure is applied to a fluid, this pressure is
        transmitted equally in all directions. In *Figure 43.1(d)*, if
        the mass of the fluid is neglected, the pressures at points
        A to D are all the same.

3   The pressure, p, at any point in a fluid depends on three
factors:

   (a) the density of the fluid, ρ in kg/m^3;
   (b) the gravitational acceleration, g, taken as 9.8 m/s^2; and
   (c) the height of fluid vertically above the point, h metres.

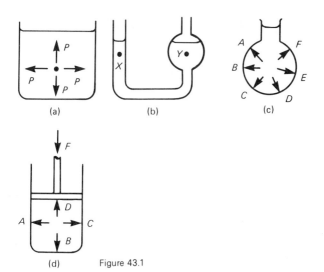

(a)   (b)   (c)

(d)   Figure 43.1

The relationship connecting these quantities is: $p = \rho g h$ **pascals**.
When the container shown in *Figure 43.2*
is filled with water of density 1000 kg/m³,
the pressure due to the water at a depth
of 0.03 m below the surface is given by:

$p = \rho g h$
$\quad = (1000 \times 9.8 \times 0.03)$
$\quad = $ **294 Pa**.

4   The air above the earth's surface is a fluid, having a density,
ρ, which varies from approximately 1.225 kg/m³ at sea level to
zero in outer space. Since $p = \rho g h$, where height h is several
thousands of metres, the air exerts a pressure on all points on the
earth's surface. This pressure, called **atmospheric pressure**, has
a value of approximately 100 kilopascals. Two terms are commonly
used when measuring pressures:

    (a) **absolute pressure**, meaning the pressure above that of
        an absolute vacuum (i.e. zero pressure), and
    (b) **gauge pressure**, meaning the pressure above that
        normally present due to the atmosphere. Thus:

**absolute pressure = atmospheric pressure
+ gauge pressure**

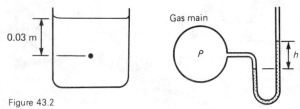

0.03 m

Figure 43.2

Gas main

P

h

Figure 43.3

Thus, a gauge pressure of 50 kPa is equivalent to an absolute pressure of $(100+50)$ kPa, i.e. 150 kPa, since the atmospheric pressure is approximately 100 kPa.

Another unit of pressure, used in particular for atmospheric pressures, is the bar.

$$1 \text{ bar} = 10^5 \text{ N/m}^2 = 100 \text{ kPa.}$$

5   There are various ways of measuring pressure, and these include by:

    (a) a U-tube manometer (see para. 6);

    (b) a barometer (see paras 7 and 8); and

    (c) a pressure gauge (see para. 9).

U-tube manometer

6   A manometer is a device used for measuring relatively small pressures, either above or below atmospheric pressure. A simple U-tube manometer is shown in *Figure 43.3*. Pressure p acting on, say, a gas main, pushes the liquid in the U-tube until equilibrium is obtained. At equilibrium: pressure in gas main, p = (atmospheric pressure, p_a) + (pressure due to the column of liquid, $\rho g h$) i.e., $p = p_a + \rho g h$.

Thus, for example, if the atmospheric pressure, p_a, is 101 kPa, the liquid in the U-tube is water of density 1000 kg/m^3 and height, h is 300 mm, then

$$\text{absolute gas pressure} = (101\,000 + 1000 \times 9.8 \times 0.3) \text{ Pa}$$
$$= (101\,000 + 2940) \text{ Pa}$$
$$103\,940 \text{ Pa} = 103.94 \text{ kPa.}$$

The gauge pressure of the gas is 2.94 kPa.

By filling the U-tube with a more dense liquid, say mercury having a density of 13 600 kg/m^3, for a given height of U-tube, the pressure which can be measured is increased by a factor of 13.6.

By inclining one limb of the U-tube, as shown in *Figure 43.4*,

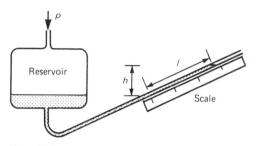

Figure 43.4

greater sensitivity is achieved, that is, there is a larger movement of the liquid for a given change in pressure when compared with a U-tube having vertical limbs. An inclined manometer normally has a reservoir of sufficient area to give virtually a constant level in the left-hand limb. From *Figure 43.4*, it can be seen that pressure p applied to the reservoir causes a scale change of 'l' in the inclined manometer compared with 'h' in a normal manometer.

Simple barometer

7   A simple barometer consists of a length of glass tubing, approximately 800 mm long and sealed at one end, which is filled with mercury and then inverted in a beaker of mercury, as shown in *Figure 43.5*. At equilibrium, the atmospheric pressure, p_a, is tending to force the mercury up the tube, whilst the force due to the column of mercury is tending to force the mercury out of the tube, i.e. $p_a = \rho gh$. As the atmospheric pressure varies, height h varies, giving an indication on the scale of the atmospheric pressure.

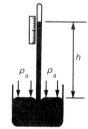

Figure 43.5

Fortin barometer

8   The Fortin barometer is as shown in *Figure 43.6*. Mercury is contained in a leather bag at the base of the mercury reservoir, and height, H, of the mercury in the reservoir can be adjusted using the screw at the base of the barometer to depress or relese the leather bag. To measure the atmospheric pressure, the screw is

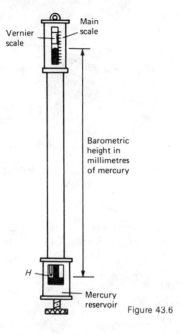

Vernier scale

Main scale

Barometric height in millimetres of mercury

H

Mercury reservoir

Figure 43.6

adjusted until the pointer at H is just touching the surface of the mercury and the height of the mercury column is then read using the main and vernier scales. The measurement of atmospheric pressure using a Fortin barometer is achieved much more accurately than by using a simple barometer.

Bourdon pressure gauge

9   The main components of a Bourdon pressure gauge are shown in *Figure 43.7*. When pressure, p, is applied to the curved phosphor bronze tube, which is sealed at A, it tends to straighten, moving A to the right. Conversely, a decrease in pressure below that due to the atmosphere moves point A to the left. When A moves to the right, B moves to the left, rotating the pointer across a scale. This type of pressure gauge can be used to measure large pressures and pressures both above and below atmospheric pressure. The Bourdon pressure gauge indicates gauge pressure and is very widely used in industry for pressure measurements.

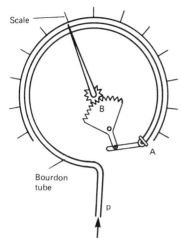

Figure 43.7

Hydrostatic pressure

10 The pressure at the base of the tank shown in *Figure 43.8(a)* is:

$$p = \rho g h = wh,$$

where w is the specific weight, i.e. the weight per unit volume, its unit being N/m³.

The pressure increases to this value uniformly from zero at the free surface. The pressure variation is shown in *Figure 43.8(b)*.

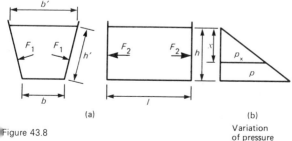

(a)

(b)

Figure 43.8

Variation of pressure

At any intermediate depth x the pressure is

$$p_x = \rho g x = wx.$$

It may be shown that the average pressure on any wetted plane surface is the pressure at the centroid, the centre of area. The sloping sides of the tank of *Figure 43.8* are rectangular and therefore the average pressure on them is the pressure at half depth:

$$\frac{p}{2} = \frac{\rho g h}{2} = \frac{wh}{2}$$

The force on a sloping side is the product of this average pressure and the area of the sloping side:

$$F_1 = \frac{\rho g h}{2} \times lh' = \frac{whlh'}{2}$$

where h' is the slant height. The pressure and consequently the forces F_1 are at right angles to the sloping sides as shown in *Figure 43.8*.

The average pressure on the vertical trapezoidal ends of the tank is not the pressure at half depth. This is because the centroid of an end is not at half depth — it is rather higher. The depth of the centroid is given by:

$$\frac{h}{3} \left(\frac{2b + b'}{b + b'} \right), \text{ (see *Figure 43.8*)}$$

The average pressure on an end is therefore:

$$\frac{\rho g h}{3} \left(\frac{2b + b'}{b + b'} \right) = \frac{wh}{3} \left(\frac{2b + b'}{b + b'} \right)$$

The forces F_2 on the vertical trapezoidal ends of the tank are horizontal forces given by the product of this average pressure and the area of the trapezium,

$$\left(\frac{b + b'}{2} \right) h.$$

The force on the base of the tank is $(\rho g h)x$ (area of base)

$$= \rho g h l b = whlb$$

For a tank with vertical sides this is the weight of liquid in the tank.

11 In any vessel containing homogeneous liquid at rest and in continuous contact, the pressure must be the same at all points at

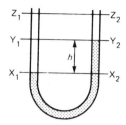

Figure 43.9

the same level. In a U-tube, as shown in *Figure 43.9*, with the liquid in the lower part at rest, the pressure must be the same on both sides for all levels up to X_1X_2. The pressure at X_1, however, is greater than the pressure at Y_1 by an amount $p_1 = w_1h = \rho_1gh$, where w_1, ρ_1 are the specific weight and density respectively of the liquid, or gas, between X_1 and Y_1.

Similarly, the pressure at X_2 is greater than the pressure at Y_2 by an amount given by $p_2 = w_2h = \rho_2gh$, where w_2, ρ_2 are the specific weight and density respectively of the liquid in the bottom of the U-tube.

For practical reasons ρ_2 must be greater than ρ_1 and the pressure at Y_1 will exceed that at Y_2 by

$$p_2 - p_1 = (w_2 - w_1)h = (\rho_2 - \rho_1)gh.$$

If the upper limits of the U-tube contain air or any other gas or gas mixture w_1 and ρ_1 can reasonably be ignored, giving

$$p_2 - p_1 = w_2h = \rho_2gh$$

If the upper limbs contain a lighter liquid, then the pressure difference may be expressed as

$$p_2 - p_1 = \rho_1(d-1)gh = (d-1)w_1h \text{ where } d = \frac{\rho_2}{\rho_1}$$

A common arrangement is mercury and water, in which case d is the relative density of mercury approximately 13.6. This gives:

$$p_2 - p_1 = 12.6\rho gh = 12.6wh,$$

ρ and w being respectively, the density and specific weight of water. The pressure difference at Z_1Z_2 will be the same as at Y_1Y_2 if both limbs contain the same liquid between these levels. This follows from the fact that the pressure increase from Z_1 to Y_1 is the same as the increase from Z_2 to Y_2.

315

Archimedes' principle

12   If a solid body is immersed in a liquid, the apparent loss of weight is equal to the weight of liquid displaced. If V is the volume of the body below the surface of the liquid, then the apparent loss of weight is $W = Vw = V\rho g$, where w, ρ are respectively the specific weight and density of the liquid.

If ρ is known and W obtained from a simple experiment, V can be calculated. Hence the density of the solid body can be calculated. If V is known and W obtained, ρ can be calculated.

If a body floats on the surface of a liquid all of its weight appears to have been lost. The weight of liquid displaced is equal to the weight of the floating body.

For example, a body weighs 2.760 N in air and 1.925 N when completely immersed in water of density 1000 kg/m³.

Hence the apparent loss of weight is

2.760 N − 1.925 N = 0.835 N.

This is the weight of water displaced, i.e. $V\rho g$, where V is the volume of the body and ρ is the density of water.

Thus 0.835 N = $(V)(1000 \text{ kg/m}^3)(9.81 \text{ m/s}^2)$

from which, volume, $V = \dfrac{0.835}{9810} = \mathbf{8.512 \times 10^{-5} \text{ m}^3}$

The density of the body = $\dfrac{\text{mass}}{\text{volume}} = \dfrac{\text{weight}}{gV}$

$$= \frac{2.760 \text{ N}}{(9.81 \text{ m/s}^2)(8.512 \times 10^{-5} \text{ m}^3)}$$

$$= \mathbf{3305 \text{ kg/m}^3}$$

Relative density = $\dfrac{\text{density}}{\text{density of water}} = \dfrac{3305}{1000} = \mathbf{3.305}$

44 Ideal gas laws

1   The relationship which exists between pressure, volume and temperature of a gas are given in a set of laws called the **gas laws**.

2   (i) **Boyle's law states:**

'*the volume V of a fixed mass of gas is inversely proportional to its absolute pressure p at constant temperature*'

i.e. $p \propto \dfrac{1}{V}$ or $p = \dfrac{k}{V}$ or $pV = k$, at constant temperature,

where

p = absolute pressure in pascals (Pa),
V = volume in m^3, and
k = a constant.

(ii) Changes which occur at constant temperature are called **isothermal** changes.

(iii) When a fixed mass of gas at constant temperature changes from pressure p_1 and volume V_1 to pressure p_2 and volume V_2 then:

$$\boxed{p_1 V_1 = p_2 V_2}$$

3   (i) **Charles' law states:**

'*for a given mass of gas at constant pressure, the volume V is directly proportional to its thermodynamic temperature T*'

i.e. $V \propto T$ or $V = kT$ or $\dfrac{V}{T} = k$, at constant pressure,

where T = thermodynamic temperature in kelvin (K).

(ii) A process which takes place at constant pressure is called an **isobaric process**.

(iii) The relationship between the Celsius scale of temperature and the thermodynamic or absolute scale is given by:

kelvin = degrees Celsius + 273

i.e. **K = °C + 273** or **°C = K − 273**

(iv) If a given mass of gas at constant pressure occupies a volume V_1 at a temperature T_1 and a volume V_2 at temperature T_2, then

$$\boxed{\frac{V_1}{T_1} = \frac{V_2}{T_2}}$$

For example, a gas occupies a volume of 1.2 litres at 20°C. If the pressure is kept constant, the volume it occupies at 130°C is determined from

$$\frac{V_1}{T_1} = \frac{V_2}{T_2},$$

i.e. $V_2 = V_1\left(\frac{T_2}{T_1}\right) = (1.2)\,\frac{(130+273)}{(20+273)}$

$$= \frac{(1.2)(403)}{(293)} = \textbf{1.65 litres}.$$

4   (i) **The Pressure law states:**

'*the pressure p of a fixed mass of gas is directly proportional to its thermodynamic temperature T at constant volume*'

i.e. $p \propto T$ or $p = kT$ or $\dfrac{P}{T} = k$

(ii) When a fixed mass of gas at constant volume changes from pressure p_1 and temperature T_1, to pressure p_2 and temperature T_2 then:

$$\boxed{\frac{p_1}{T_1} = \frac{p_2}{T_2}}$$

5   (i) **Dalton's law of partial pressure states:**

'*the total pressure of a mixture of gases occupying a given volume is equal to the sum of the pressures of each gas, considered separately, at constant temperature*'.

(ii) The pressure of each constituent gas when occupying a fixed volume alone is known as the **partial pressure** of that gas.

6   An **ideal gas** is one which completely obeys the gas laws given in paras 2 to 5. In practice no gas is an ideal gas, although air is very close to being one. For calculation purposes the difference between an ideal and an actual gas is very small.

7   (i) Frequently, when a gas is undergoing some change, the

pressure, temperature and volume all vary simultaneously. Provided there is no change in the mass of a gas, the above gas laws can be combined giving:

$$\frac{p_1 V_1}{T_1} = \frac{p_2 V_2}{T_2} = k, \quad \text{where } k \text{ is a constant.}$$

(ii) For an ideal gas, constant $k = mR$, where m is the mass of the gas in kg, and R is the **characteristic gas constant**,

$$\text{i.e. } \frac{pV}{T} = mR \text{ or } \boxed{pV = mRT}$$

This is called the characteristic gas equation. In this equation, p = absolute pressure in pascals, V = volume in m^3, m = mass in kg, R = characteristic gas constant in J/(kg K) and T = thermodynamic temperature in kelvin.

(iii) Some typical values of the characteristic gas constant R include: air, 287 J/(kg K), hydrogen 4160 J/(kg K), oxygen 260 J/(kg K) and carbon dioxide 184 J/(kg K). For example, some air at a temperature of 40°C and pressure 4 bar occupies a volume of 0.05 m^3. The mass of air is determined from

$$pV = mRT$$

Hence, mass $m = \dfrac{pV}{RT} = \dfrac{(4 \times 10^5 \text{ Pa})(0.05 \text{ m}^3)}{(287 \text{ J/(kg K)})(40 + 273) \text{ K}}$

$$= \textbf{0.223 kg or 223 g}.$$

8 **Standard temperature and pressure (i.e. STP)** refers to a temperature of 0°C, i.e. 273 K, and normal atmospheric pressure of 101.325 kPa.

Kinetic theory of gases

9 The kinetic theory of gases suggests that gases are composed of particles in motion. The continual bombardment of any surface by the gas causes a pressure to be exerted; the greater the density of a gas, the more frequent the number of collisions between molecules and the surface and the greater the pressure exerted. Hence the pressure increases either when the volume of a certain mass of gas is reduced, or when more gas is pumped into a vessel. When the temperature of a gas is increased, the speed of the molecules increases, causing an increase in both the number and the momentum imparted by each collision. This accounts for the increase in pressure of a gas with increase in temperature.

Maxwell (in 1860) explained some of the properties of a gas by assuming that the molecules of a gas make elastic collisions, spend negligible time actually in collision, and themselves occupy a negligible part of the volume of the gas. Also, the attractive forces between molecules are assumed negligible.

10   It may be shown that for a gas occupying a volume V at pressure p and containing n molecules each of mass m moving at an average velocity of c,

$$pV = \tfrac{1}{3}mnc^2$$

Also, the kinetic energy of the molecules of a gas is proportional to its thermodynamic temperature.

11   When a liquid evaporates molecules with sufficient kinetic energy escape from the liquid's surface. The higher the temperature of the liquid the greater the average kinetic energy of the molecules and the greater the number of molecules which are able to escape. Since it is the molecules with the highest kinetic energy which escape the average kinetic energy of the remaining molecules decreases and thus the liquid cools.

12   If a liquid evaporates a **vapour** is formed. When a vapour exists in the presence of its own liquid a **saturated vapour** is formed. If all the liquid evaporates an **unsaturated vapour** iis produced. The higher the temperature the greater the number of molecules which escape to form the vapour. These molecules bombard the walls of the container and thus exert a pressure.

The **saturated vapour pressure** depends only on the

Table 44.1

Temperature ($°C$)	Saturated vapour pressure of water ($10^3\ Pa$)
0	0.61
10	1.23
20	2.33
30	4.23
40	7.35
50	12.3
60	19.9
70	31.2
80	47.4
90	70.2
100	101
150	476
200	1550

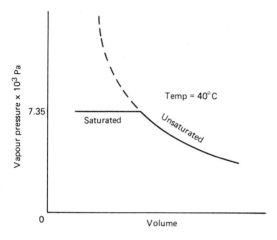

Figure 44.1

temperature of the vapour. The saturated vapour pressure of water at various temperatures is shown in *Table 44.1*.

A liquid boils at a temperature when its saturated vapour pressure is equal to the atmospheric pressure. Thus water will boil at a temperature greater than 100°C if the atmospheric pressure is increased. This is the principle of the pressure cooker.

13   A saturated vapour does not obey the gas laws since its pressure depends only on temperature. An unsaturated vapour will obey the gas laws fairly closely as long as it remains unsaturated. If an unsaturated vapour at a particular temperature is decreased in volume its pressure will rise in accordance with Boyle's law until it reaches the saturated vapour pressure at that particular temperature (see *Figure 44.1*). When the vapour pressure at 40°C reaches 7.35×10^3 Pa the vapour becomes saturated as it starts to liquify.

45 **Properties of water and steam**

1   When two systems are at different temperatures, the transfer of energy from one system to the other is called **heat transfer**. For a block of hot metal cooling in air, heat is transferred from the hot metal to the cool air. The **principle of conservation of energy** may be stated as

'*energy cannot be created nor can it be destroyed*',

and since heat is a form of energy, this law applies to heat transfer problems. A more convenient way of expressing this law when referring to heat transfer problems is:

initial energy of the system + energy entering the system
   = final energy of the system + energy leaving the system

or,

energy entering the system
   = change of energy within the system
      + energy leaving the system

2   Fluids consist of a very large number of molecules moving in random directions within the fluid. When the fluid is heated, the speeds of the molecules are increased, increasing the kinetic energy of the molecules. There is also an increase in volume due to an increase in the average distance between molecules, causing the potential energy of the fluid to increase. The **internal energy**, U, of a fluid is the sum of the internal kinetic and potential energies of the molecules of a fluid, measured in joules. It is not usual to state the internal energy of a fluid as a particular value in heat transfer problems, since it is normally only the **change** in internal energy which is required.

3   The sum of the internal energy and the pressure energy of a fluid is called the **enthalpy** of the fluid, denoted by the symbol H and measured in joules. The pressure energy, or work done, is given by the product of pressure, p, and volume V, that is: pressure energy = pV joules. Thus,

enthalpy = internal energy + pressure energy (or work done), i.e.

$$H = U + pV.$$

As for internal energy, the actual value of enthalpy is usually unimportant and it is the **change** in enthalpy which is usually required. In heat transfer problems involving steam and water, water is considered to have zero enthalpy at a standard pressure of 101 kPa and a temperature of 0°C. The word 'specific' associated with quantities indicates 'per unit mass'. Thus the **specific enthalpy** is obtained by dividing the enthalpy by the mass and is denoted by the symbol h.

Thus **specific enthalpy** $= \dfrac{\text{enthalpy}}{\text{mass}} = \dfrac{H}{m} = h$

The units of specific enthalpy are joules per kilogram (J/kg).

4   The specific enthalpy of water, h_f, at temperature $\theta°C$ is the quantity of heat needed to raise 1 kg of water from 0°C to $\theta°C$, and is called the **sensible heat** of the water. Its value is given by:

specific heat capacity of water $(c) \times$ temperature change (θ)

i.e. $h_f = c\theta$

The specific heat capacity of water varies with temperature and pressure but is normally taken as 4.2 kJ/kg, thus

$h_f = 4.2\ \theta\ \text{kJ/kg.}$

5   When water is heated at a uniform rate, a stage is reached (at 100°C at standard atmospheric pressure), where the addition of more heat does not result in a corresponding increase in temperature. The temperature at which this occurs is called the **saturation temperature**, t_{SAT}, and the water is called **saturated water**. As heat is added to saturated water, it is turned into **saturated steam**. The amount of heat required to turn 1 kg of saturated water into saturated steam is called the **specific latent heat of vaporisation**, and is given the symbol, h_{fg}. The total specific enthalpy of steam at saturation temperature, h_g, is given by:

the specific sensible heat + the specific latent heat of vaporisation,

i.e. $h_g = h_f + h_{fg}.$

6   If the amount of heat added to saturated water is insufficient to turn all the water into steam, then the ratio

$\dfrac{\text{mass of saturated steam}}{\text{total mass of steam and water}}$ is called the **dryness fraction**

of the steam, denoted by the symbol q. The steam is called **wet steam** and its total enthalpy is given by:

enthalpy of saturated water + (dryness fraction)

$\times$ (enthalpy of latent heat of vaporisation)

i.e. $h_f + q h_{fg}$.

7 When the amount of heat added to water at saturation temperature is sufficient to turn all the water into steam, it is called either saturated vapour or **dry saturated steam**. The addition of further heat results in the temperature of the steam rising and it is then called **superheated steam**. The specific enthalpy of superheated steam above that of dry saturated steam is given by

$c(t_{SUP} - t_{SAT})$

where c is the specific heat capacity of the steam and t_{SUP} is the temperature of the superheated steam. The total specific enthalpy of the superheated steam is given by

$h_f + h_{fg} + c(t_{SUP} - t_{SAT})$,

or $h_g + c(t_{SUP} - t_{SAT})$.

8 The relationship between temperature and specific enthalpy can be shown graphically and a typical temperature-specific enthalpy diagram is shown in *Figure 45.1*. In this figure, AB represents the sensible heat region where any increase in enthalpy results in a corresponding increase in temperature. BC is called the **evaporation line** and points between B and C represent the wet steam region, point C representing dry saturated steam. Points to the right of C represent the superheated steam region.

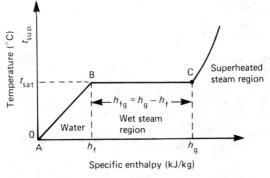

Figure 45.1

9   The boiling point of water, t_{SAT} and the various specific enthalpies associated with water and steam (h_f, h_{fg}, h_g and $c(t_{SUP} - t_{SAT})$), all vary with pressure. These values at various pressures have been tabulated in **steam tables**, extracts from these being shown in *Tables 45.1* and *45.2*.

In *Table 45.1*, the pressure, in both bar and kilopascals, and saturated water temperature are shown in columns on the left. The

Table 45.1

Pressure		Saturation temperature t_{SAT} (°C)	Specific enthalpy (kJ/kg)		
(bar)	(kPa)		Saturated Water h_f	Latent heat h_{fg}	Saturated Vapour h_g
1	100	99.6	417	2258	2675
1.5	150	111.4	467	2226	2693
2	200	120.2	505	2202	2707
3	300	133.5	561	2164	2725
4	400	143.6	605	2134	2739
5	500	151.8	640	2109	2749
6	600	158.8	670	2087	2757
7	700	165.0	697	2067	2764
8	800	170.4	721	2048	2769
9	900	175.4	743	2031	2774
10	1000	179.9	763	2015	2778
15	1500	198.3	845	1947	2792
20	2000	212.4	909	1890	2799
30	3000	233.8	1008	1795	2803
40	4000	250.3	1087	1714	2801

Table 45.2

Pressure		Saturation temperature t_{SAT} (°C)	Saturated vapour h_g	Specific enthalpy (kJ/kg) Superheated steam at				
(bar)	(kPa)			200°C	250°C	300°C	350°C	400°C
1	100	99.6	2675	2876	2975	3075	3176	3278
1.5	150	111.4	2693	2873	2973	3073	3175	3277
2	200	120.2	2707	2871	2971	3072	3174	3277
3	300	133.5	2725	2866	2968	3070	3172	3275
4	400	143.6	2739	2862	2965	3067	3170	3274
5	500	151.8	2749	2857	2962	3065	3168	3272
6	600	158.8	2757	2851	2958	3062	3166	3270
7	700	165.0	2764	2846	2955	3060	3164	3269
8	800	170.4	2769	2840	2951	3057	3162	3267
9	900	175.4	2774	2835	2948	3055	3160	3266
10	1000	179.9	2778	2829	2944	3052	3158	3264
15	1500	198.3	2792	2796	2925	3039	3148	3256
20	2000	212.4	2799		2904	3025	3138	3248
30	3000	233.8	2803		2858	2995	3117	3231
40	4000	250.3	2801			2963	3094	3214

columns on the right give the corresponding specific enthalpies of water, (h_f) and dry saturated steam (h_g), together with the specific enthalpy of the latent heat of vaporisation (h_{fg}).

The columns on the right of *Table 45.2* give the specific enthalpies of dry saturated steam, (h_g) and superheated steam at various temperatures. The values stated refer to zero enthalpy. However, if the degree of superheat is given, this refers to the saturation temperature. Thus at a pressure of 100 kPa, the column headed, say, 250°C has a degree of superheat of $(250-99.6)$°C, that is, 150.4°C.

For example, let some dry saturated steam at a pressure of 1.0 MPa be cooled at constant pressure until it has a dryness fraction of 0.6. The change in the specific enthalpy of the steam is determined as follows:

From *Table 45.1*, the specific enthalpy of dry saturated steam h_g at a pressure of 1.0 MPa (i.e. 1000 kPa) is 2778 kJ/kg.
From para. 6, the specific enthalpy of wet steam is $h_f + q h_{fg}$.
At a pressure of 1.0 MPa, h_f is 763 kJ/kg and h_{fg} is 2015 kJ/kg.

Thus the specific enthalpy of the wet steam $= 763 + 0.6 \times 2015$
$$= 1972 \text{ kJ/kg}.$$

The change in the specific enthalpy is $2778 - 1972 = \textbf{806 kJ/kg}$.

As another example, let steam leave a boiler at a pressure of 3.0 MPa and a temperature of 400°C. The degree of superheat may be determined from *Table 45.2*.

At a pressure of 3.0 MPa, i.e. 3000 kPa, the saturation temperature is 233.8°C, hence the degree of superheat is $400-233.8=\textbf{166.2°C}$. The specific enthalpy of superheated steam at 3.0 MPa and 400°C is given in *Table 45.2* as **3231 kJ/kg**.

10   Superheated steam behaves very nearly as if it is an ideal gas and the gas laws introduced in Chapter 44 may be used to determine the relationship between pressure, volume and temperature.

46 **Fluids in motion**

Bernoulli's equation

1   This is the principle of the conservation of energy applied to fluids in motion.

$$\frac{p}{w}+\frac{v^2}{2g}+Z=\textbf{constant} \qquad (1)$$

or $$\frac{p}{\rho}+\frac{v^2}{2}+Zg=\textbf{constant} \qquad (2)$$

All of the quantities on the left of each equation apply to a specified fixed point in the moving fluid:

   p=pressure (gauge pressure unless otherwise specified)
   w=specific weight (weight per unit volume)
   v=velocity
   g=acceleration due to gravity
   Z=height above some specified datum
   ρ=density

The two significant differences between this application of the principle of the conservation of energy and the application of the principle to solids in motion are:
(i) this application is to a steady process mass (or weight) flowing per second has to be considered instead of a given fixed mass or weight;
(ii) a third form of energy, that is, pressure energy, must be considered; the corresponding form of energy in dealing with solids, strain energy, is only occasionally met with.
Each of the terms in equation (1) represents energy per unit weight of fluid. The basic unit of each term is the metre,

   i.e. $\dfrac{\text{Nm}}{\text{N}}$ or $\dfrac{\text{J}}{\text{N}}$.

The basic unit for pressure is the same as for stress. N/m^2 or pascal (Pa). Each term in the equation is called a **head**.

$\dfrac{p}{w}$ is the **pressure head** $\left(\dfrac{N/m^2}{N/m^3} = m \right)$

$\dfrac{v^2}{2g}$ is the **velocity or kinetic head** $\left(\dfrac{(m/s)^2}{m/s^2} = m \right)$

z is the **potential head** (m).

The sum of the three heads is called **total head** (**H**).
Equation (2) gives energies per unit mass.

In practice, as with solids, some energy is lost, converted into heat. The estimation of this loss is an important aspect of the mechanics of fluids at a higher level.

. If a pipe is filled by moving liquid the volumetric rate of flow $\dot{V}$, that is, the volume passing per second, must be the same at every section: $V = A_1 v_1 = A_2 v_2$ where A_1 and v_1 are the cross-sectional area and velocity at one selected section and A_2 and v_2 are area and velocity at a second section. The equation $A_1 v_1 = A_2 v_2$ is called the equation of continuity.

The basic unit of volumetric rate of flow $\dot{V}$ is m^3/s which is a large unit. The smaller unit, litre per second, is often preferred.

$$1 \text{ litre} = 1000 \text{ cm}^3 = 1 \text{ m}^3 \times 10^{-3}$$

(For very low rates of flow litres per minute units may be preferred.)

For example, let the rate of flow of water through a pipe of 32 mm diameter be 2.8 litres/s. The total head at a point where the pressure is 28.4 kPa with reference to a datum 1.84 m below is determined as follows:

Velocity of flow $v = \dfrac{\dot{V}}{A} = \dfrac{2.8 \times 10^{-3} \text{ m}^3/\text{s}}{\dfrac{\pi}{4}(32)^2 \times 10^{-6} \text{ m}^2} = 3.4815 \text{ m/s}.$

The velocity head, $\dfrac{v^2}{2g} = \dfrac{(3.4815)^2}{(2)(9.81)} = 0.6178 \text{ m}.$

The specific weight

$w = \rho g = (1000 \text{ kg/m}^3)(9.81 \text{ m/s}^2) = 9.81 \text{ kN/m}^3.$

The pressure head, $\dfrac{p}{w} = \dfrac{28.4 \text{ kPa}}{9.81 \text{ kN/m}^2} = 2.895 \text{ m}.$

The potential head $Z = 1.84$ m.

Hence the total head $H = \dfrac{P}{w} + \dfrac{v^2}{2g} + Z$

$$= 2.895 + 0.6178 + 1.84 = \textbf{5.353 m}.$$

Flow through orifices

2 Water issuing from a tank as a horizontal jet, as shown in *Figure 46.1*, has a velocity head only, if the datum is taken at the level of orifice. Water which will eventually form the jet starts at

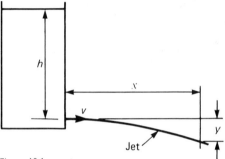

Figure 46.1

the top of the tank with a potential head only, h. Equating initial potential and final velocity heads,

$$\frac{v^2}{2g} = h$$

$$v = \sqrt{(2gh)}$$

In practice, some energy loss occurs. The ratio

$$\frac{\text{actual velocity of jet}}{\text{theoretical velocity of jet}}$$

is called the **coefficient of velocity** of the orifice (C_v). The actual velocity can be obtained from accurate observation of co-ordinates x and y of points on the jet trajectory.

 The diameter of the jet is also found in practice to be less than the diameter of the orifice. The ratio:

$$\frac{\text{cross-sectional area of jet}}{\text{area of orifice}}$$

is called the **coefficient of contraction** of the orifice (C_c). The ratio

$$\frac{\text{actual rate of discharge}}{\text{theoretical rate of discharge}}$$

is called the **coefficient of discharge** of the orifice (C_d). But

$$\frac{\text{actual rate of discharge}}{\text{theoretical rate of discharge}}$$

$$= \frac{\text{actual velocity of jet} \times \text{c.s.a. of jet}}{\text{theoretical velocity of jet} \times \text{area of orifice}}$$

i.e. $\boldsymbol{C_d = C_v \times C_c}$.

For example, an orifice in the bottom of a water tank has a diameter of 12.5 mm. Assuming coefficients of contraction and velocity of 0.64 and 0.96 respectively, the depth of water required in the tank to give a rate of discharge through the orifice of 0.25 litres per second is determined as follows:

The theoretical rate of discharge from the tank,

$$\dot{\mathbf{v}} = Av = \left(\frac{\pi d^2}{4}\right) \sqrt{(2gh)}$$

$$= \frac{\pi (0.0125)^2}{4} \sqrt{[(2)(9.81)h]}$$

$$= 0.00054357 \sqrt{h} \text{ m}^3/\text{s}$$

$$= 0.54357 \sqrt{h} \text{ litres/s, where } h \text{ is in metres.}$$

Coefficient of discharge $C_d = C_v \times C_c = 0.96 \times 0.64 = 0.6144$.

$$C_d = \frac{\text{actual rate of discharge}}{\text{theoretical rate of discharge}},$$

thus (actual rate of discharge) = (C_d) (theoretical rate of discharge)

$$= (0.6144)(0.54357 \sqrt{h})$$

$$= 0.33397 \sqrt{h} \text{ litres/s.}$$

Hence, $0.25 = 0.33397 \sqrt{h}$,
from which, depth of water

$$h = \left(\frac{0.25}{0.33397} \right)^2 = 0.5604 \text{ m} = \mathbf{560.4 \text{ mm}}.$$

Impact of a jet

3   The force exerted by a jet of water on a plate is, from
Newton's third law of motion, equal and opposite to the force
exerted by the plate on the water. From Newton's second law this
is equal to the rate of change of momentum of water.

$$\text{Rate of change of momentum} = \frac{\text{mass} \times \text{change of velocity}}{\text{time}}$$

In dealing with solids this is interpreted as

$$\text{mass} \times \frac{\text{change of velocity}}{\text{time}} = \text{mass} \times \text{acceleration}$$

In dealing with the continuous process of fluid flow it must be
interpreted as

$$\frac{\text{mass}}{\text{time}} \times \text{change of velocity}$$

$$= \text{mass rate of flow} \times \text{change of velocity } (\dot{M}v).$$

In the case of the jet striking a flat
plate at right angles (see *Figure 46.2*) the
final velocity in the original direction is
zero, so that v is the change of velocity in
this direction. Also, if d is the diameter of
the jet:

$$\dot{M} = \dot{V}p = Av\rho$$

$$= \frac{\pi d^2}{4} v\rho$$

Force on plate,

$$\mathbf{F = \dot{M}v = \frac{\pi d^2}{4} v^2 \rho.}$$

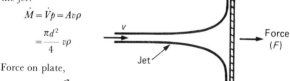

Figure 46.2

The force will be in newtons if the jet diameter is in metres,
the jet velocity in metres per second and density in kilograms per
metre cubed (mass rate of flow $\dot{M}$ in kg/s).

For example, let a jet of water with a diameter of 12.5 mm
and a velocity of 40 m/s strike a stationary flat plate at right
angles.

Mass rate of flow

$$\dot{M} = Av\rho = \frac{\pi d^2}{4} v\rho = \frac{\pi (0.0125)^2}{4} (40)(1000)$$

$$= 4.9087 \text{ kg/s}.$$

Force on plate $F = \dot{M}v = (4.9087 \text{ kg/s})(40 \text{ m/s}) = 196.3 \text{ kg m/s}^2$
$$= \textbf{196.3 N}.$$

47 Measurement of fluid flow

1   The measurement of fluid flow is of great importance in many industrial processes, some examples including air flow in the ventilating ducts of a coal mine, the flow rate of water in a condenser at a power station, the flow rate of liquids in chemical processes, the control and monitoring of the fuel, lubricating and cooling fluids of ships and aircraft engines, and so on.

   Fluid flow is one of the most difficult of industrial measurements to carry out, since flow behaviour depends on a great many variables concerning the physical properties of a fluid.

2   There are available a large number of fluid flow measuring instruments, generally called **flow meters**, which can measure the flow rate of liquids (in m^3/s) or the mass flow rate of gaseous fluids (in kg/s). The two main categories of flow meters are differential pressure flow meters and mechanical flow meters.

3   (i) When certain flow meters are installed in pipelines they often cause an obstruction to the fluid flowing in the pipe by reducing the cross-sectional area of the pipeline. This causes a change in the velocity of the fluid, with a related change in pressure. *Figure 47.1* shows a section through a pipeline into which a flow meter has been inserted. The flow rate of the fluid may be determined from a measurement of the difference between the pressures on the walls of the pipe at specified distances upstream

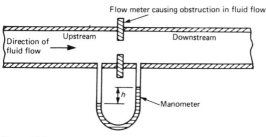

Figure 47.1

and downstream of the flowmeter. Such devices are known as **differential pressure flow meters**.

(ii) The pressure in *Figure 47.1* is measured using a manometer connected to appropriate pressure tapping points. The pressure is seen to be greater upstream of the flow meter than downstream, the pressure difference being shown as *h*. Calibration of the manometer depends on the shape of the obstruction, the positions of the pressure tapping points and the physical properties of the fluid.

(iii) Examples of differential pressure flow meters commonly used include:

     (a) Orifice plate (see para. 6).
     (b) Venturi tube (see para. 7),
     (c) Flow nozzles (see para. 8), and
     (d) Pitot-static tube (see para. 9).

(iv) British standard reference BS 1042: Part 1: 1964 and Part 2A: 1973 'Methods for the measurement of fluid flow in pipes' give specifications for measurement, manufacture, tolerances, accuracy, sizes, choice, and so on, of differential flow meters.

4   (i) With mechanised flow meters a sensing element situated in a pipeline is displaced by the fluid flowing past it.

(ii) Examples of mechanical flow meters commonly used include:

     (a) Deflecting vane flow meter (see para. 10), and
     (b) Turbine type meters (see para. 11).

5   Other flow meters available include:

     (a) Float and tapered-tube meter (see para. 12),
     (b) Electromagnetic flow meter (see para. 13), and
     (c) Hot wire anemometers (see para. 14).

Orifice plate

6   (a) An orifice plate consists of a circular, thin, flat plate with a hole (or orifice) machined through its centre to fine limits of accuracy. The orifice has a diameter less than the pipeline into which the plate is installed and a typical section of an installation is shown in *Figure 47.2(a)*. Orifice plates are manufactured in stainless steel, monel metal, polyester glass fibre, and for large pipes, such as sewers or hot gas mains, in brick and concrete.

     When a fluid moves through a restriction in a pipe, the fluid accelerates and a reduction in pressure occurs, the magnitude of which is related to the flow rate of the fluid. The variation of pressure near an orifice plate is shown in *Figure 47.2(b)*. The position of minimum

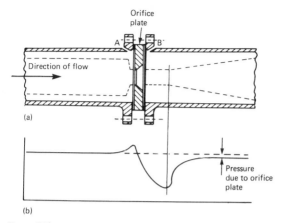

Orifice plate

A B

Direction of flow

(a)

(b)

Pressure due to orifice plate

Figure 47.2

pressure is located downstream from the orifice plate where the flow stream is narrowest. This point of minimum cross-sectional area of the jet is called the '*vena contracta*'. Beyond this point the pressure rises but does not return to the original upstream value and there is a permanent pressure loss. This loss depends on the size and type of orifice plate, the positions of the upstream and downstream pressure tappings and the change in fluid velocity between the pressure tappings which depends on the flow rate and the dimensions of the orifice plate.

In *Figure 47.2(a)* corner pressure tappings are shown at A and B. Alternatively, with an orifice plate inserted into a pipeline of diameter d, pressure tappings are often located at distances of d and $d/2$ from the plate respectively upstream and downstream. At distance d upstream the flow pattern is not influenced by the presence of the orifice plate and distance $d/2$ coincides with the vena contracta.

(b) **Advantages of orifice plates are:**

    (i) they are relatively inexpensive; and

    (ii) they are usually thin enough to fit between an existing pair of pipe flanges.

(c) **Disadvantages of orifice plates are:**

    (i) the sharpness of the edge of the orifice can become

worn with use, causing calibration errors;

(ii) the possible build-up of matter against the plate; and

(iii) a considerable loss in the pumping efficiency due to the pressure loss downstream of the plate.

Orifice plates are usually used in medium and large pipes and are best suited to the indication and control of essentially constant flow rates. Several applications are found in the general process industries.

Venturi tube

Construction

7   (a) The Venturi tube or venturimeter is an instrument for measuring with accuracy the flow rate of fluids in pipes. A typical arrangement of a section through such a device is shown in *Figure 47.3*, and consists of a short converging

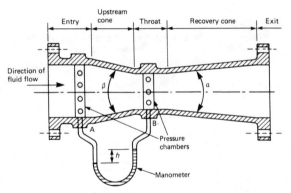

Figure 47.3

conical tube called the inlet or upstream cone leading to a cylindrical portion called the throat. This is followed by a diverging section called the outlet or recovery cone. The entrance and exit diameter is the same as that of the pipeline into which it is installed. Angle β is usually a maximum of $21°$, giving a taper of $\beta/2$ of $10\frac{1}{2}°$. The length of the throat is made equal to the diameter of the throat. Angle α is about $5°$ to $7°$ to ensure a minimum loss of energy but where this is unimportant α can be as large as $14°$ or $15°$.

Pressure tappings are made at the entry (at A) and at the throat (at B) and the pressure difference h which is measured using a manometer or similar gauge, is dependant on the flow rate through the meter. Usually pressure chambers are fitted around the entrance pipe and the throat circumference with a series of tapping holes made in the chamber to which the manometer is connected. This ensures an average pressure is recorded. The loss of energy due to turbulence which occurs just downstream with an orifice plate is largely avoided in the venturimeter due to the gradual divergence beyond the throat.

Venturimeters are usually made a permanent installation in a pipeline and are usually manufactured from stainless steel, cast iron, monel metal or polyester glass fibre.

(b) **Advantages of venturimeters:**
(i) High accuracy results are possible.
(ii) There is a low pressure loss in the tube (typically only 2%–3% in a well proportioned tube).
(iii) Venturimeters are unlikely to trap any matter from the fluid being metered.

(c) **Disadvantages of venturimeters:**
(i) High manufacturing cost.
(ii) The installation tends to be rather long (typically 120 mm for a pipe of internal diameter 50 mm).

Flow nozzles

8 The flow nozzle lies between the orifice plate and the venturimeter both in performance and cost. A typical section through a flow nozzle is shown in *Figure 47.4* where pressure tappings are located immediately adjacent to the upstream and downstream faces of the nozzle (i.e. at points A and B). The fluid does not contract any further as it leaves the nozzle and the pressure loss created is considerably less than that occurring with orifice plates.

Flow nozzles are suitable for use with high velocity flows for they do not suffer the wear that occurs in orifice plate edges during such flows.

Pitot-static tube

9 (a) A Pitot-static tube is a device for measuring the velocity of moving fluids or of the velocity of bodies moving

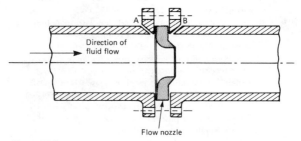

Figure 47.4

through fluids. It consists of one tube, called the Pitot tube, with an open end facing the direction of the fluid motion, shown as pipe R in *Figure 47.5* and a second tube, called the static tube, with the opening at 90° to the fluid flow, shown as T in *Figure 47.5*. Pressure is recorded by a pressure gauge moving with the flow, i.e. static or stationary relative to the fluid. This is called static pressure and connecting a pressure gauge to a small hole in the wall of a pipe, such as point T in *Figure 47.5*, is the easiest method of recording this pressure. The difference in pressure $(p_R - p_T)$, shown as h in the manometer of *Figure 47.5* is an indication of the speed of the fluid in the pipe.

Figure 47.6 shows a practical Pitot-static tube consisting of a pair of concentric tubes. The centre tube is the impact probe which has an open end which faces 'head-on' into the flow. The outer tube has a series of holes around its circumference located at right angles to the flow — as seen by AB in *Figure 47.6*. The manometer, showing a pressure difference of h, may be calibrated to indicate the velocity of flow directly.

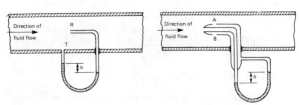

Figure 47.5

Figure 47.6

Applications

(b) A Pitot-static tube may be used for both turbulent and non-turbulent flow. The tubes can be made very small compared with the size of the pipeline and the monitoring of flow velocity at particular points in the cross-section of a duct can be achieved. The device is generally unsuitable for routine measurements and in industry is often used for making preliminary tests of flow rate in order to specify permanent flow measuring equipment for a pipeline. Another use is the measurement of ventilating-duct air flows in mines.

However, the main use of Pitot tubes is to measure the velocity of solid bodies moving through fluids, such as the velocity of ships and of low speed aircraft. In these cases, the tube is connected to a Bourdon pressure gauge which can be calibrated to read velocity directly. A development of the Pitot tube, a pitometer, tests the flow of water in water mains and detects leakages.

(c) **Advantages of Pitot-static tubes:**
   (i) They are inexpensive devices;
   (ii) They are easy to install;
   (iii) They produce only a small pressure loss in the tube;
   (iv) They do not interrupt the flow.

(d) **Disadvantages of Pitot-static tubes:**
   (i) Due to the small pressure difference, they are only suitable for high velocity fluids;
   (ii) They can measure the flow rate only at a particular position in the cross-section of the pipe;
   (iii) They easily become blocked when used with fluids carrying particles.

Deflecting vane flow meter

10   The deflecting vane flow meter consists basically of a pivoted vane suspended in the fluid flow stream as shown in *Figure 47.7*.

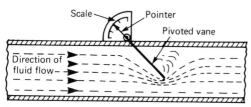

Figure 47.7

When a jet of fluid impinges on the vane it deflects from its normal position by an amount proportional to the flow rate. The movement of the vane is indicated on a scale which may be calibrated in flow units. This type of meter is normally used for measuring liquid flow rates in open channels or for measuring the velocity of air in ventilation ducts. The main disadvantages of this device are that it restricts the flow rate and it needs to be recalibrated for fluids or differing densities.

Turbine type flow meters

11 Turbine type flow meters are those which use some form of multi-vane rotor and are driven by the fluid being investigated. Three such devices are the cup anemometer, the rotary vane positive displacement meter and the turbine flow meter.

Cup anemometer
(a) An anemometer is an instrument which measures the velocity of moving gases and is most often used for the measurement of wind speed. The cup anemometer has 3 or 4 cups of hemispherical shape mounted at the end of arms radiating horizontally from a fixed point. The cup system spins round the vertical axis with a speed approximately proportional to the velocity of the wind. With the aid of a mechanical and/or electrical counter the wind speed can be determined and the device is easily adapted for automatic recording.

(b) **Rotary vane positive displacement meters** measure the flow rate by indicating the quantity of liquid flowing through the meter in a given time. A typical device is shown in section in *Figure 47.8* and consists of a cylindrical chamber into which is placed a rotor containing a number of vanes (six in this case). Liquid entering the chamber turns the rotor and a known

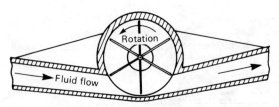

Figure 47.8

amount of liquid is trapped and carried round to the outlet.

If x is the volume displaced by one blade then for each revolution of the rotor in *Figure 47.8*, the total volume displaced is $6x$. The rotor shaft may be coupled to a mechanical counter and electrical devices which may be calibrated to give flow volume. This type of meter in its various forms is widely used for the measurement of domestic and industrial water consumption, for the accurate measurement of petrol in petrol pumps and for the consumption and batch control measurements in the general process and food industries for measuring flows as varied as solvents, tar and mollases (i.e. thickish treacle).

(c) A **turbine flow meter** contains in its construction a rotor to which blades are attached which spin at a velocity proportional to the velocity of the fluid which flows through the meter. A typical section through such a meter is shown in *Figure 47.9*. The number of revolutions made by the turbine blades may be determined by a mechanical or electrical device enabling the flow rate or total flow to be determined.

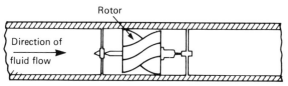

Figure 47.9

Advantages of turbine flow meters include a compact durable form, high accuracy, wide temperature and pressure capability and good response characteristics. Applications include the volumetric measurement of both crude and refined petroleum products in pipelines up to 600 mm bore, and in the water, power, aerospace, process and food industries, and with modification may be used for natural, industrial and liquid gas measurements. Turbine flow meters require periodic inspection and cleaning of the working parts.

Float and tapered-tube flow meter

Construction

(a) With orifice plates and venturimeters the area of the opening in the obstruction is fixed and any change in the flow rate produces a corresponding change in pressure. With the float and tapered tube meter the area of the restriction may be varied so as to maintain a steady pressure differential. A typical meter of this is shown diagrammatically in *Figure 47.10* where a vertical tapered tube contains a 'float' which has a density greater than the fluid.

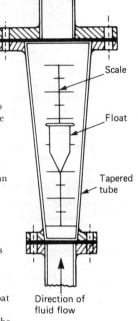

The float in the tapered tube produces a restriction to the fluid flow. The fluid can only pass in the annular area between the float and the walls of the tube. This reduction in area produces an increase in velocity and hence a pressure difference, which causes the float to rise. The greater the flow rate, the greater is the rise in the float position, and vice-versa.

Direction of fluid flow

Figure 47.10

The position of the float is a measure of the flow rate of the fluid and this is shown on a vertical scale engraved on a transparent tube of plastic or glass. For air, a small sphere is used for the float but for liquids there is a tendency to instability and the float is then designed with vanes which cause it to spin and thus stabilise itself as the liquid flows past. Such meters are often called 'rotameters'. Calibration of float and tapered tube flow meters can be achieved using a Pitot-static tube or by installing an orifice plate or venturimeter in the pipeline.

(b) **Advantages of float and tapered tube flow meters:**

  (i) Has a very simple design;

      (ii) Can be made direct reading;

      (iii) Can measure very low flow rates.

(c) **Disadvantages of float and tapered tube flow meters:**

      (i) They are prone to errors, such as those caused by temperature fluctuations;

      (ii) They can only be installed vertically in a pipeline;

      (iii) They cannot be used with liquids containing large amounts of solids in suspension;

      (iv) They need to be recalibrated for fluids of different densities.

(d) **Practical applications of float and tapered tube meters** are found in the medical field, in instrument purging, in mechanical engineering test rigs and in simple process applications, in particular for very low flow rates. Many corrosive fluids can be handled with this device without complications.

Electromagnetic flow meter

13 (a) The flow rate of fluids which conduct electricity, such as water or molten metal, can be measured using an electromagnetic flow meter whose principle of operation is based on the laws of electromagnetic induction. When a conductor of length l moves at right angles to a magnetic field of density B at a velocity of v, an induced e.m.f. e is generated, given by $e = Blv$ (see Chapter 12).

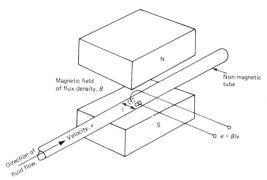

Figure 47.11

With the electromagnetic flow meter arrangement shown in *Figure 47.11*, the fluid is the conductor and the e.m.f. is detected by two electrodes placed across the diameter of the non-magnetic tube. Rearranging $e = Blv$ gives

velocity $v = \dfrac{e}{Bl}$

Thus with B and l known, when e is measured, the velocity of the fluid can be calculated.

(b) **Main advantages of electromagnetic flow meters:**
(i) Unlike other methods, there is nothing to directly impede the fluid flow;
(ii) There is a linear relationship between the fluid flow and the induced e.m.f.;
(iii) Flow can be metered in either direction by using a centre zero measuring instrument.

(c) **Applications of electromagnetic flow meters** are found in the measurement of speeds of slurries, pastes and viscous liquids, and are also widely used in the water production, supply and treatment industry.

Hot-wire anemometer

(a) A simple hot wire anemometer consists of a small piece of wire which is heated by an electric current and positioned in the air or gas stream whose velocity is to be measured. The stream passing the wire tends to cool it, the rate of cooling being dependant on the flow velocity. There are various ways in which this is achieved:
(i) If a constant current is passed through the wire, variation in flow results in a change of temperature of the wire and hence a change in resistance which may be measured by a Wheatstone bridge arrangement. The change in resistance may be related to fluid flow.
(ii) If the wire's resistance, and hence temperature, is kept constant, a change in fluid flow results in a corresponding change in current which can be calibrated as an indication of the flow rate.
(iii) A thermocouple may be incorporated in the assembly monitoring the hot wire and recording the temperature which indicates the air or gas velocity.

(b) **Advantages of the hot-wire anemometer:**
It is small and has great sensitivity.

48 Simple harmonic motion and natural vibrations

Simple harmonic motion

1   Simple harmonic motion is defined as a periodic motion of a point along a straight line, such that its acceleration is always towards a fixed point in that line and is proportional to its distance from that point.

Simple harmonic motion (SHM) may be considered as the projection on a diameter of a movement at uniform speed around the circumference of a circle.

In *Figure 48.1* P moves with uniform speed v ($=\omega r$) around a circle of radius r; the point X projected from P on diameter AB moves with SHM. The acceleration of P is the centripetal acceleration, $\omega^2 r$. The displacement (measured from the mean position O), the velocity and the acceleration of X are respectively:

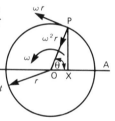

Figure 48.1

displacement, $x = OX = r\cos r = \cos \omega t$
(where t is the time measured from the instant when P and X are at A and $\theta = 0$)
velocity, $v = \omega r\sin\theta = -\omega r\sin\omega t$
acceleration, $a = -\omega^2 r\cos\theta$
$\qquad = -\omega^2 r\cos\omega t$
$\qquad = -\omega^2 x$.

The expressions for velocity and acceleration can be derived from that for displacement by differentiating with respect to time. The negative signs in the expressions for velocity and acceleration show that for the position of X in *Figure 48.1* both velocity and acceleration are in the opposite direction from the displacement. Displacement and acceleration are always in opposite directions. The periodic time T of the motion is the time taken for one complete oscillation of X. In this time OP makes one complete revolution and therefore:

periodic time $T = \dfrac{2\pi}{\omega}$

Since $a = \omega^2 x$, $\omega = \sqrt{\left(\dfrac{a}{x}\right)}$

and therefore

$$T = 2\pi \sqrt{\left(\dfrac{x}{a}\right)} = 2\pi \sqrt{\left(\dfrac{\text{displacement}}{\text{corresponding acceleration}}\right)}$$

The frequency n is the number of complete oscillations in one second.

$$n = \dfrac{\omega}{2\pi} = \dfrac{1}{T} = \dfrac{1}{2\pi} \sqrt{\left(\dfrac{a}{x}\right)}$$

The unit for n of one oscillation per second is called the hertz (Hz). The maximum velocity of X occurs at the mid-point, where it equals the velocity of P.

$v_{max} = \omega r$

The maximum acceleration of X occurs at the extreme positions A and B where it equals the acceleration of P:

$a_{max} = \omega^2 r$.

The velocity of X is zero at A and B; its acceleration is zero at 0. The amplitude of the oscillation is r. The distance AB ($2r$) is sometimes called the stroke or travel of the motion.

For example, a body moves with SHM of amplitude 45 mm and frequency 2.5 Hz.

Hence, frequency $= \dfrac{\omega}{2\pi} = 2.5$,

from which, $\omega = (2\pi)(2.5) = 5\pi = 15.708$ rad/s.

The maximum velocity $= \omega r = 15.708 \times 45 = $ **706.9 mm/s** and this occurs at the mean position.

The maximum acceleration $= \omega^2 r = (15.708)^2 \times 45 = $ **11.10 m/s^2**, which occurs at each extreme position.

For a displacement of, say, 25 mm, $\cos \theta = \frac{25}{45}$ (see *Figure 48.1*) from which

$\theta = 56.25°$

The velocity for a displacement of 25 mm from the mean position is $\omega r \sin \theta$, i.e. $(15.708)(45)(0.8315) = $ **587.7 mm/s**.

The acceleration for a displacement of 25 mm from the mean

position is $\omega^2 r \cos \theta = \omega^2 x$, i.e. $(15.708)^2(25) =$ **6169 mm/s²** or **6.169 m/s²**.

Natural vibration

Motion closely approximating to SHM occurs in a number of natural or free vibrations. Many examples are met with where a body oscillates under a control which obeys Hooke's law, for example a spring or a beam. Consider, for example, the helical spring shown in *Figure 48.2*. If, from its position of rest, the mass M is pulled down a distance r and then released, the mass will oscillate in a vertical line. In the rest position the force in the spring will exactly balance the force of gravity acting on the mass.

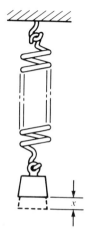

If s is the stiffness of the spring, that is force per unit change of length, then for a displacement x from the rest position, the change in the force in the spring is sx. This change of force is the unbalanced or accelerating force F acting on the mass M.

Figure 48.2

$$F = sx$$

Acceleration, $a = \dfrac{F}{M} = \dfrac{s}{M} x$

This shows that the acceleration is directly proportional to displacement from its rest position. The motion is therefore SHM. The periodic time is given by:

$$T = 2\pi \sqrt{\left(\frac{x}{a}\right)} = 2\pi \sqrt{\left(\frac{x}{\dfrac{sx}{M}}\right)} = 2\pi \sqrt{\left(\frac{M}{s}\right)}$$

For example, a load of 10 kg is hung from a vertical helical spring and it causes an extension of 15 mm. The load is pulled down a further distance of 18 mm and then released.

Thus, the weight of the load $= Mg = 10 \times 9.81 = 98.1$ N

Stiffness of the spring,

$$s = \frac{\text{force}}{\text{extension}} = \frac{98.1 \text{ N}}{15 \text{ mm}} = 6.54 \text{ N/mm} = 6.54 \text{ kN/m}.$$

Frequency of the vibration

$$n = \frac{1}{T} = \frac{1}{2\pi\sqrt{\left(\dfrac{M}{s}\right)}} = \frac{\sqrt{s}}{2\pi\sqrt{M}} = \frac{\sqrt{(6.54 \text{ kN/m})}}{2\pi\sqrt{(10 \text{ kg})}}$$

$$= \frac{\sqrt{\left(654 \dfrac{\text{kg m/s}^2}{\text{kg m}}\right)}}{2\pi} = \mathbf{4.07 \text{ Hz}}.$$

The amplitude r of the vibration is 18 mm. The maximum velocity of the load is

$$\omega r = (2\pi n)(r) = (2\pi 4.07)(18) = \mathbf{460.3 \text{ mm/s}}.$$

The maximum acceleration of the load is

$$\omega^2 r = (2\pi 4.07)^2 (18) = \mathbf{11\,771 \text{ mm/s}^2}$$

$$\text{or } \mathbf{11.77 \text{ m/s}^2}.$$

The maximum force in the spring
= maximum extension × stiffness
= $(15 \text{ mm} + 18 \text{ mm})(6.54 \text{ N/mm}) = \mathbf{215.8 \text{ N}}$.

Another common example of a vibration giving a close approximation to SHM is the movement of a **simple pendulum**. This is defined as a mass of negligible dimensions on the end of a cord or rod of negligible mass. For a small displacement x of the bob A from its mean position C (*Figure 48.3*), the accelerating force F on the bob, weight W, is $W \sin \theta$, which equals $W\theta$ very nearly if θ is a small angle and measured in radians.
The acceleration of the bob is therefore

$$a = \frac{F}{M} = \frac{W\theta}{M} = \frac{Mg\theta}{M} = g\theta$$

where M is the mass of the bob, $= W/g$.
But $x = l\theta$ when θ is measured in radians,

Hence, $\theta = \dfrac{x}{l}$ and $a = g\theta = \dfrac{g}{l}x$

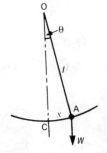

Figure 48.3

The acceleration is therefore shown to be proportional to

displacement, satisfying the definition of SHM and the periodic time is:

$$T = 2\pi \sqrt{\left(\frac{x}{a}\right)} = 2\pi \sqrt{\left(\frac{l}{g}\right)}$$

For example, a simple pendulum has a length of 780 mm. Periodic time of the pendulum

$$T = 2\pi \sqrt{\left(\frac{l}{g}\right)} = 2\pi \sqrt{\left(\frac{0.78 \text{ m}}{9.81 \text{ m/s}^2}\right)} = \textbf{1.772 s}.$$

If the amplitude of movement of the bob is 80 mm, then the maximum velocity of the bob is

$$\omega r = \left(\frac{2\pi}{T}\right) r,$$

where r is the amplitude.

Hence maximum velocity $= \left(\dfrac{2\pi}{1.772}\right)(80) = \textbf{283.8 mm/s}$,

and maximum acceleration $= \omega^2 r = \left(\dfrac{2\pi}{T}\right)^2 r$

$$= \left(\frac{2\pi}{1.772}\right)^2 (80)$$

$$= \textbf{1006 mm/s}^2 \text{ or } \textbf{1.006 m/s}^2.$$

The angular motion of the pendulum must not be confused with the angular motion of an imaginary line used in the analysis of simple harmonic motion. The imaginary line (OP in *Figure 47.1*) rotates at a constant speed. The angular velocity of the pendulum is variable, having its maximum value in the vertical position. For a velocity v of the bob, the angular velocity of the pendulum is

$$\omega_{\text{p}} = \frac{v}{l}$$

The angular acceleration of the pendulum is greatest in the extreme positions. For an acceleration a of the bob, the angular acceleration of the pendulum is

$$\alpha_{\text{p}} = \frac{a}{l}.$$

Index

355

Butterworths Technician Series